I0503970

TABLE OF CONTENTS

ACRONYMS

USCG	United States Coast Guard
USG	United State Government
EEZ	Exclusive Economic Zone
UNCLOS	United Nations Convention on the Law of the Sea

ILLUSTRATIONS

INTRODUCTION

> We, the people, still believe that our obligations as Americans are not just to ourselves, but to all posterity. We will respond to the threat of climate change, knowing that the failure to do so would betray our children and future generations.
> — President Obama, *Inaugural Address*, January 21, 2013

> The ice breaker is a trouble shooter in Polar navigation.
> — LTC Joseph R. Russ, US Army Command and General Staff College, 1947

Our Iceberg is Melting is a book by John Kotter and Holger Rathgeber about adapting to changes in a dynamic world.[1] In the book, penguins find themselves living on an iceberg that is melting and as a group must overcome the challenges of change and embrace the opportunities presented to them or face extinction. Today, the Earth's northern iceberg is melting and like the penguins, the nations of the world must find ways to adapt. Scientific evidence states that the Arctic is experiencing an unprecedented climate change.[2] The temperature in the Arctic continues to increase eight times faster than that of the rest of the globe, which in turn brings increased precipitation at one percent per decade. The extent of the sea ice continues to decrease by almost three percent each decade, with summer decreases at almost seven and a half percent. Additionally, sea ice thickness is also getting thinner. In the summer of 2010, sea ice thickness was seventy percent below the 1979 figures. Research revealed that in 2011 the Arctic sea ice extent was the lowest ever recorded.[3] Most importantly, current projections have ice-free summers in the Arctic by 2030.[4] This tremendous change in the Arctic climate has far-reaching impacts on every nation

---

[1] John Kotter and Holger Rathgeber, *Our Iceberg is Melting* (New York: St Martin's Press).

[2] This paper uses the US definition of the Arctic as per the Arctic Research and Policy Act of 1984 (15 U.S.C. 4111). The Arctic is that region which encompasses all US and foreign territory north of the Arctic Circle and all US territory north and west of the boundary formed by the Porcupine, Yukon, and Kuskokwim Rivers, and all contiguous seas and straits north of and adjacent to the Arctic Circle.

[3] Georg Heygster, *"Arctic Sea Ice Extent Small as Never Before"*, University of Bremen, http://www.iup.uni-bremen.de:8084/amsr/minimum2011-en.pdf (accessed 17 January 2013); Also see figure 1.

[4] Ice-free is a misleading term. Ice-free is described best as areas of open water in the ice that is navigable by ships. Mr. Mark Meza of the US Coast Guard describes this environment as "navigable

and on Earth as a whole.

The recent changes in the Arctic create numerous areas of interest for the United States, largely driven by national security and economic goals. In light of these recent environmental changes around the globe, there has been an increase in human activity in the Arctic. Commercial shipping companies seeking shorter routes have begun transiting the Arctic along with cruise ships and scientific research vessels. The potential for large revenues from untapped oil, gas, and minerals has energy companies conducting exploration. In addition, increased interest in the Arctic's marine fisheries and mammals has nations, conservation groups, and commercial fishermen venturing to the region. The large potential for additional natural resources also has nation states increasing research activity, staking claims, and asserting their sovereignty in the Arctic.

Although the United States has multiple interests in the Arctic, and numerous departments and agencies have developed their own Arctic strategies, the United States has failed to put forth the effort and funding necessary to get on top of the emerging changes in the Arctic. The USG has not satisfactorily implemented its national strategy as laid out by President George W. Bush in the January 9, 2009 National Security Presidential Directive 66 - Homeland Security Presidential Directive 25 (NSPD 66/HSPD 25), nor has it developed additional or updated Executive level guidance. [5] NSPD 66/HSPD 25 provides seven primary areas of concern: national and homeland security, international governance, continental shelf and boundary issues, promotion of international scientific cooperation, Arctic maritime transportation, economic issues, and environmental protection and conservation of natural resources. However, until 2013, the US Congress failed to provide sufficient funding to implement NSPD 66/HSPD 25. Most notably, the United States does not have the requisite surface ships and icebreakers capable of

---

Arctic with ice infested waters". U.S. Navy, *Arctic Environmental Assessment and Outlook Report; In support of the Navy Arctic Roadmap; Action Item 5.7* (Washington, DC, 2011). v.

[5] President, *National Security Presidential Directive/NSPD 66 – Homeland Security Presidential Directive /HSPD 25* (January 9, 2009). For the text of NSPD 66/HSPD 25, see Appendix A.

supporting and defending US interests in the Arctic. Furthermore, given sequestration and a continuing resolution, which does not allow the start of new projects, even the 2013 funding is in jeopardy.

It has been four years since President Bush issued NSPD 66/HSPD 25. A lot has changed during those four years; in fact, President Barack Obama took office and won reelection for a second term, but still has not published any additional guidance on the Arctic except for a short sentence in the National Security Strategy, which reaffirms President Bush's policy:

> The United States is an Arctic Nation with broad and fundamental interests in the Arctic region, where we seek to meet our national security needs, protect the environment, responsibly manage resources, account for indigenous communities, support scientific research, and strengthen international cooperation on a wide range of issues.[6]

Yet the Arctic continues to melt, presenting numerous opportunities and challenges. An appreciation of the US national interests and an overview of background issues in the Arctic serve as a starting point to examine the US Government (USG) actions, or lack thereof, regarding the changes in the Arctic. A summary of the USG policy on the Arctic will provide a general understanding of the fundamental principles the President laid out in the USG Policy. Additionally, multiple government agencies continue to provide insights into the US national interests and policy. And, although the United States has a national policy in place regarding the Arctic, the USG is a large bureaucracy of so many different departments and agencies that there are numerous government policies regarding the Arctic. In addition, independent, USG departments and agencies have issued guidance such as the United States Coast Guard's (USCG) Arctic Strategic Approach[7] and the US Navy's Arctic Road Map.[8] Examining these policies will shed light on how America has not adequately implemented its Arctic policy. Finally, at detailed assessment of actions taken by USG departments and agencies in the seven areas laid out in the

---

[6] President, *National Security Strategy 2010* (May 2010), 50.

[7] U.S. Coast Guard. Commandant Instruction 16003.1, *U.S. Coast Guard Arctic Strategic Approach* (Washington, DC, April 26, 2011).

[8] U.S. Department of the Navy, *U.S. Navy Arctic Roadmap* (Washington, DC, October 2009).

USG policy, reveals poor overall implementation of policy. The US should implement NSPD 66 – HSPD 25 immediately in order to preserve its interests in the Arctic. The US also needs additional icebreakers in order to conduct more research, project power and assert sovereignty, gain Arctic domain awareness, ensure safety of Arctic maritime shipping and to assist in building partnerships with other nations.

US INTERESTS

National Security and opportunities for economic growth are the two major factors that drive all US interests in the Arctic. National and Homeland Security remain the US top priorities. With the opening of the Arctic, national borders have become more easily accessible, causing a need for all Arctic nations to assert sovereignty and protect their borders. For the US, this means there is now easier access to the mainland US through both Alaska and Canada, which is an ever-growing concern. In order to protect these vulnerable borders the US must have the military and border protection capabilities to operate in the harsh Arctic environment. For the US military, that currently means missile defense and early warning, maritime domain awareness, and access to and freedom of navigation, under and over the seas. For US Customs and Border Patrol "[t]he high volume of commerce and travel between the United States and Canada creates opportunities for criminals to conceal their cross-border activity. The potential for terrorists or violent extremists to attempt an attack or gain entry across the land, air, or maritime borders poses the single greatest security threat along the border." [9]

Economic opportunities, however, drive the majority of the US interests in the Arctic. At the forefront is oil and gas. The US Energy Information Administration projected that the Arctic holds thirteen percent of the world's oil reserves and thirty percent of its undiscovered gases. [10] In September 2012, Shell Oil began drilling for oil in the U.S. Chukchi Sea off the North Coast of Alaska. Shell spent

---

[9] U.S. Department of Homeland Security. *Northern Border Strategy* (Washington, DC, June 2012), 6.

[10] Heather Conley, Terry Toland, and Jamie Kraut, *A New Security Architecture for the Arctic: An American Perspective* (Washington, DC: CSIS, January 2012), 2.

$4.5 billion on Arctic Ocean drilling before even beginning to drill in September including $2.8 billion to the US government for leases in the Chukchi and Beaufort Seas. [11] But, after the 2012 drilling season, Shell developed a major setback when one of its rigs ran aground on a remote island just north of Kodiak, Alaska and on February 27, 2013 Shell Oil confirmed that it will not drill in US Arctic waters during the 2013 season due to damage suffered to the drill rig in the grounding. The halt in drilling is only temporary according to Shell Oil President Marvin Odum who stated, "Shell remains committed to building an Arctic exploration program that provides confidence to stakeholders and regulators, and meets the high standards the company applies to its operations around the world." [12]

The Arctic is also rich with inorganic resources to include nickel, iron ore, and other natural minerals. One-fifth of the world's nickel comes from mines in the Russian Arctic. The Russian Arctic also produces half of the global supply of palladium, which is widely used in catalytic converters and in dentistry. [13] Canada estimates that in 2014 it will begin shipping eighteen million tons of iron ore per year from Baffinland with operations expected to run for at least twenty-five years. [14] The US currently has six large mineral mines operating in Alaska. Alaska's Red Dog mine is the worlds' largest supplier of zinc, producing roughly ten percent of the global supply. [15]

---

[11] Dan Joling, Associated Press, "Shell begins oil, gas drilling off Alaska coast," *USA Today* September 10, 2012, http://usatoday30.usatoday.com/money/business/story/2012/09/10/shell-begins-oil-gas-drilling-off-alaska-coast/57720768/1(accessed Mar 9, 2013); Also see figure 1.

[12] Associated Press, "Report says Shell unprepared for Arctic drilling," *St. Louis Post-Dispatch* March 9, 2012, http://www.stltoday.com/news/national/shell-suspends-drilling-for-arctic-ocean-in/article_adfd3e50-249b-59d9-92bf-091daf400d8c.html (accessed Mar 9, 2013).

[13] Heather Conley, Terry Toland, and Jamie Kraut. *A New Security Architecture for the Arctic: An American Perspective.* (Washington, DC: CSIS, January 2012), 5; and, Northwest Territorial Mint, "Palladium Uses," http://bullion.nwtmint.com/palladium_uses.php (accessed April 1, 2013).

[14] David Tinsley, Lloyd's List, "Arctic gold rush drives evolution of ice vessels," Who Owns the Arctic? Blog, entry posted January 22, 2009, http://byers.typepad.com/arctic/2009/01/arctic-gold-rush-drives-evolution-of-ice-vessels.html (accessed March 9, 2013).

[15] Alaska Department of Natural Resources. *Alaska Mineral Industry 2010: Special Report 65*, by D.J. Szumigala, L.A. Harbo, and J.N. Adleman (Fairbanks: Alaska Department of Natural Resources,

Another issue that holds great economic potential is the opening of shipping routes through the

Arctic. "With the opening of trade routes across the Arctic Ocean from the North Atlantic to the North

Pacific, trillion-dollar business opportunities will alter the global balance of power- as other new trade

routes have before them."[16] Three possible maritime transit routes may open up to commercial maritime

traffic as the Arctic ice melts, thereby shortening transit times and resulting in major cost savings to

suppliers and shipping companies.[17] The first, the Northern Sea Route, that links Europe and Asia via

Russian Federation waters, is already widely in use. The second is the Northwest Passage, an international

strait, through Canadian Arctic waters.[18] The third is a route through US waters of the Bering Strait, into

the Arctic and across the North Pole then past Greenland to the North Atlantic Ocean.[19]

An example of the benefits of these shipping routes occurred in 2011, when thirty-four cargo

ships carrying 820,000 tons of cargo used the Northern Sea Route. Escorted by Russian Federation

nuclear icebreakers, the ships cut the transit time from Asia to Europe and North American by one third

as opposed to transiting through the Suez Canal.[20] All three of these newly opened sea routes would

greatly increase the maritime traffic transiting through the Bering Strait between Alaska and the Russian

Federation, which could bring greater economic prosperity to Alaska's Aleutian Islands. The Northern

---

Division of Geological and Geophysical Surveys, 2011), 6.

[16] Paul Arthur Berkman, *Environmental Security in the Arctic Ocean, Promoting Co-operation and Preventing Conflict* (Abingdon OX14 4RN, UK: Royal United Service Institute for Defence and Security Studies (RUSI)), 4.

[17] Aldo Chircop, "The Growth of International Shipping in the Arctic: Is a Regulatory Review Timely?" *The International Journal of Marine and Coastal Law*, 24, no. 2 (2009), 355-380.

[18] The US claims that the Northwest Passage is an international strait in accordance with UNCLOS; however, Canada claims the Northwest Passage as internal Canadian waters.

[19] Aldo Chircop, "The Growth of International Shipping in the Arctic: Is a Regulatory Review Timely?" *The International Journal of Marine and Coastal Law,* 24, no. 2 (2009), 355-380; See Figure 1.

[20] Heather Conley, Terry Toland, and Jamie Kraut. *A New Security Architecture for the Arctic: An American Perspective* (Washington, DC: CSIS, 2012), 8.

Sea Route would reduce a maritime distance between East Asia (Japan) and Western Europe from 11,200 miles using the Suez Canal to 6,500 miles, cutting transit time by more than forty percent. The Northwest Passage maritime trip between Seattle and Western Europe would take only 7,000 miles, rather than the 9,000 miles by using the Panama Canal.[21]

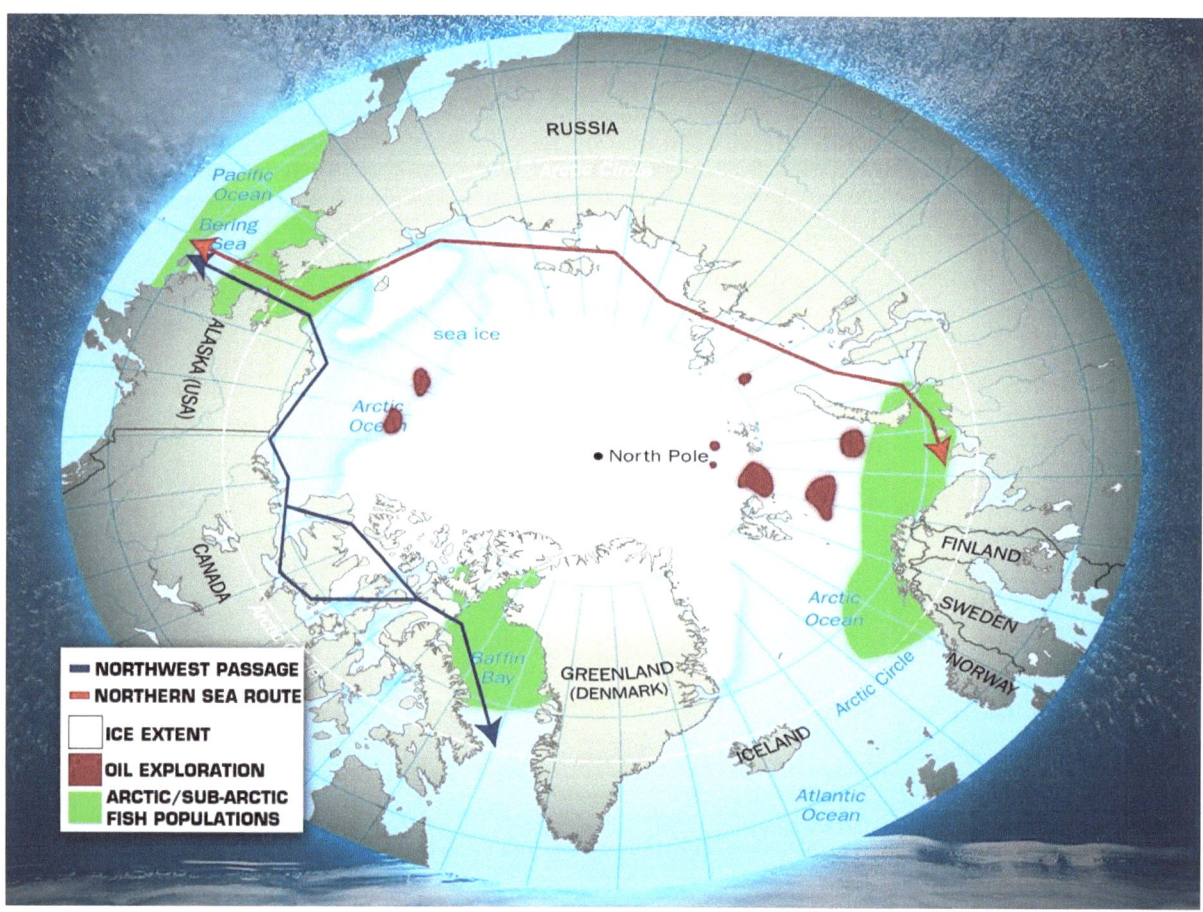

Figure 1. A chart depicting the Arctic region at minimum sea ice extent, in September 2012, shipment routes, ice extent, oil exploration, and fish populations.

Source: Capt. Jonathan S. Spaner, USCG, Coast Guard director, Office of Emerging Policy.

The Arctic also has an abundance of Living Marine Resources, both fish and marine mammals.[22]

---

[21] Scott G. Borgerson, "Arctic Meltdown: The Economic and Security Implications of Global Warming," *Foreign Affairs*, vol 87, no. 2. (March/April 2008), 69; See figure 1.

[22] See Figure 1.

Currently, Bering Sea fisheries make up sixty percent of the US commercial fishing industry.[23] In 2009, the US banned commercial fishing in a 150,000 square nautical mile section of US Arctic waters as a precaution pending further research of the ecosystem in the region.[24] The ban includes an area under dispute with Canada, bringing additional pressures on the two nations to develop a diplomatic resolution to the US and Canada border disputes.[25] In the US, whales, seals, walruses, and polar bears are protected under the Marine Mammal Protection Act. The taking of these mammals in US waters or by US citizens on the high seas, including the Arctic is illegal.[26] However, the US needs to conduct additional research to determine how climate change will affect these and other mammals that live in the Arctic and how in turn that will affect the Arctic's indigenous population.

The Arctic is home to numerous indigenous people groups. All but one Arctic nation has indigenous citizens, almost all of whom are minorities. A large number of indigenous people rely on traditional hunting, fishing, and herding which are being threatened by the climate change in the Arctic. Additionally, many of these people live with either their own form of government of participate within their nation's governmental framework, making up the majority in their respective districts, such as in Canada and Alaska's Northern Slope. Others have unique representation in their nation's governments, like that of the Saami parliaments in Norway, Finland, and Sweden. The melting of the Arctic also presents an increasing threat of rising water replacing the land these people have lived on for centuries,

---

[23] President, Proclamation, "Fisheries of the United States Exclusive Economic Zone Off Alaska; Fisheries of the Arctic Management Area; Bering Sea Subarea," *Federal Register* 74, no. 211 (November 3, 2009), 56734.

[24] Lawson W. Brigham, Capt, USCG (ret), "The Fast-Changing Maritime Arctic," *Proceedings*, U.S. Naval Institute (May 2010), 58.

[25] Marport, Happenings from the World of Deep Sea Technology, "United Stated Arctic Fishing Policy Latest in Can-Am Dispute," September 3, 2009, http://blog.marport.com/2009/09/03/1329/ (accessed March 9, 2013).

[26] *Marine Mammal Protection Act (MMPA)*, U.S. Code 16§1361.

while the increase in industrialization brings modern wage jobs to the area, also displacing the native way of life. As the Arctic warms there is an increased risk to the health of the local population including insect and wildlife borne illnesses, and exposure to additional pollutants such as mercury. To understand better these opportunities and risks, many indigenous people have formed international groups, such as the Inuit Circumpolar Council form by the Inuit people of the U.S., Canada, Greenland, and Russia. The Inuit Circumpolar Council successfully petitioned the 2007 United Nations General Assembly to adopt the Declaration on the Rights of Indigenous Peoples. And, in 2009 the U.N. mandated a greater role in indigenous people on matters of climate change. [27]

BACKGROUND INFORMATION

The United Nations Convention on the Law of the Sea (UNCLOS) is the primary international law in the Arctic. UNCLOS is an international treaty that became effective in November 1994, which established laws for operating on, over, and under the seas. It institutes rules regarding international maritime borders, territorial seas, exclusive economic zones (EEZ), continental shelf jurisdiction, deep seabed mining, ship construction, and laws in international waters.[28] UNCLOS was forwarded to the US Senate for ratification on October 6, 1994, but the Senate has yet to ratify it.[29] Under Article 76 of UNCLOS, nations may file a claim for jurisdiction of natural resources, including oil and gas, of its continental shelf.[30] Without being signatory to UNCLOS, the US cannot file a continental shelf claim nor

---

[27] Congressional Research Service, *Changes in the Arctic: Background and Issues for Congress* by Ronald O'Rourke. Washington, DC, 2013. 31-35.

[28] United Nations Convention on the Law of the Sea (UNCLOS), December 10, 1982, *United Nation Treaty Series*, volume 1833, registration Number I-31363.

[29] Senate Committee on Foreign Relations, *The Law of the Sea Convention (Treaty Doc.103-39): The U.S. National Security and Strategic Imperatives for Ratification.* 112th Cong., 2d sess., 2012.

[30] Under UNCLOS, a nation has ten years after becoming a member to file a continental shelf claim.

does it have the right to view other nations claims; however, all claims have thus far been made public.[31]

In September 1996, the eight Arctic nations as well as multiple Arctic indigenous people, organizations joined to form the Arctic Council. [32] The Arctic Council was formed as a non-directive body to "promote cooperation, coordination and interaction" with a particular goal of "sustainable development and environmental protection" of the Arctic. As such, the Arctic Council does not address issues of military security.[33] The Arctic Council is a high-level forum to discuss issues and come to agreements on issues using diplomacy, but it is not legally binding. As part of its organizational structure, the Arctic Council has several working groups: Arctic Monitoring and Assessment Program (AMAP), Protection of the Arctic Marine Environment (PAME), Arctic Contaminants Action Program (ACAP), Conservation of Arctic Fauna and Flora (CAFF), Emergency Prevention, Preparedness and Response (EPPR), Sustainable Development Working Group (SDWG). The Arctic Council has conducted seven major studies and published corresponding reports. They include reports on the Arctic Environment, Radioactivity, Climate Impact, Human Development, Oil and Gas, Human Health and Arctic Marine Shipping. [34]

In addition to the Arctic Council, some individuals and groups desire an Arctic Treaty much like that of the Antarctic Treaty, which states that the Antarctic is to be used for peaceful purposes only.[35]

---

[31] Congressional Research Service, *Changes in the Arctic: Background and Issues for Congress* by Ronald O'Rourke. Washington, DC, 2011, 11.

[32] The eight Arctic Nations are: United States, Canada, Denmark, Finland, Iceland, Norway, Russia, and Sweden. The Arctic indigenous people organizations are: Inuit Circumpolar Conference, Saami Council and Association of the Indigenous Minorities of the North, Siberia and the Far East of the Russian Federation.

[33] Arctic Council, *Declaration on the Establishment of the Arctic Council* (Ottawa, 1996), 1.

[34] Paul Arthur Berkman, *Environmental Security in the Arctic Ocean, Promoting Co-operation and Preventing Conflict* (Abingdon OX14 4RN, UK: Royal United Service Institute for Defence and Security Studies (RUSI)), 58-59.

[35] Ibid., 63-64.

However, members of the Arctic Council, including the US, agree that the UNCLOS is the legal controlling body and that it satisfactorily covers all aspects of the Arctic.[36] Therefore, the current overarching international agreement that governs the Arctic remains UNCLOS, as it provides guidance for everything pertaining to the sea including: territorial sea and contiguous zone, straits used for international navigation, archipelagic states, exclusive economic zone, continental shelf, high seas, regime of islands, protection and preservation of the marine environment, marine scientific research.

In the Arctic, however, there remain disputed areas, especially regarding internal waters and straits, which may cause international jurisdictional conflict.[37] Regarding ice-covered areas UNCLOS specifically states,

> Coastal States have the right to adopt and enforce non-discriminatory laws and regulations for the prevention, reduction and control of marine pollution from vessels in ice-covered areas within the limits of the exclusive economic zone, where particularly severe climatic conditions and the presence of ice covering such areas for most of the year create obstructions or exceptional hazards to navigation, and pollution of the marine environment could cause major harm to or irreversible disturbance of the ecological balance. Such laws and regulations shall have due regard to navigation and the protection and preservation of the marine environment based on the best available scientific evidence.[38]

To date the US has not implemented any specific regulations using this law; however, in July 2010 Canada began enforcing a new Northern Canada Vessel Traffic Services, known as NORDREG CANADA. It is mandatory for ships 300 tons or greater or vessels carrying dangerous cargos to report in ninety-six hours prior to entering Canadian Arctic waters.[39]

A third international governing body is the International Maritime Organization (IMO) created in 1948 as an agency of the United Nations. IMO is responsible for the oversight of safety and security of

---

[36] Arctic Council, *The Ilulissat Declaration* (Ilulissat, Greenland, 2008), 1.

[37] US cases will be discussed later in this chapter.

[38] United Nations Convention on the Law of the Sea, December 10, 1982, *United Nation Treaty Series*, volume 1833, registration Number I-31363.

[39] Canada, Minister of Fisheries and Oceans Canada. Canadian Coast Guard. *Ice Navigation in Canadian Waters, 2012*. (Ottawa, Ontario, 2012), 3-4.

shipping as well as the prevention of ship pollution.[40] Since the Exxon Valdez grounding and resulting oil spill in Alaska, IMO has been working to establish international standards for vessels operating in the Arctic. Known as the Polar Code, these standards lay out construction design specifications for ships operating in different ice conditions. However, as with anything international, consensus is hard to obtain, so currently these Polar Codes are only guidelines.

Another very important international agreement that affects the Arctic is The *Convention on the International Regulations for Preventing Collisions at Sea, 1972* (COLREGS). The COLREGS is the international navigation rules of the road to prevent collisions on the high seas.[41] In the US, there are two sets of Navigation rules, one for international waters and one for inland US waters. Fortunately, all Arctic waters of the US fall within international waters, the same as the high seas. COLREGS apply to Arctic shipping; however, they do not specifically address ships transiting thru ice. Although according to the rules a vessel engaged in ice breaking could claim to be "restricted in its ability to maneuver" which would give them right of way over typical power driven vessels, this does not include vessels operating in or near ice other than ships actually engaged in icebreaking.[42]

Even with an international framework in the Arctic, there still exist multiple demands for border protection and projection of power and sovereignty. There are several international territorial disputes in the Arctic, of which the US is directly involved in three. The US claims that the Northwest Passage through the Canadian northern archipelago is an international strait per UNCLOS. However, Canada claims it to be an inland waterway and therefore sovereign Canadian waters.[43] In 1995, to maintain its

---

[40] International Maritime Organization, "Introduction to IMO," http://www.imo.org/About/Pages/Default.aspx (accessed January 3, 2013).

[41] Arctic Council, *Arctic Marine Shipping Assessment 2009 Report*, April 2009.

[42] International Maritime Organization, *Convention on the International Regulations for Preventing Collisions at Sea, 1972*, codified at 33 USCS § 1602.

[43] Congressional Research Service, *Changes in the Arctic: Background and Issues for Congress* by Ronald O'Rourke. Washington, DC, 2013, 11.

claim, the US sent USCGC POLAR SEA through the Northwest Passage without requesting permission from the Canadian Government. The Canadian government responded by making a straight baseline claim under UNCLOS, essentially making an international statement that the Northwest Passage is Canadian internal waters. [44] Following this incident the US and Canada, although still in dispute, did sign an agreement that all USCG cutters would notify Canada before transiting the strait. [45] The US and Canada also have a boundary dispute in the Beaufort Sea off the north coast of Alaska, which has already caused some friction over fisheries management and may potentially cause conflict over other natural resources. The third US dispute is with the Russian Federation in the Bering Sea. However, both parties follow an agreement signed in 1990, which the US Senate has ratified, but the Russian Federation Duma has not yet approved. [46]

Although the US has not filed an extended continental shelf claim in the Arctic, other nations have. Most notorious is the Russian Federation's claim of the Lomonosov Ridge that extends under the North Pole. The Russian Federation has made substantiating the Lomonosov Ridge as part of their continental shelf a top strategic priority and has even placed a Russian Federation flag on the sea floor underneath the North Pole. [47] There have been concerns, that if the Russian claim is not accepted by UNCLOS, then the Russian Federation would unilaterally claim the Lomonosov Ridge, which is in concert with US assertion of continental shelf rights prior to UNCLOS. [48] Since the US has not ratified

---

[44] Luke R. Petersen, "International Strait or Internal Waters?: The navigational potential of the Northwest Passage," *USCG Proceedings* (Summer 2009), 45.

[45] "Canada and United States of America Agreement on Arctic Cooperation," January 11, 1988, *United Nations Treaty Series*, no. 1852-i-31529 (Ottawa, 1988).

[46] Congressional Research Service, *Changes in the Arctic: Background and Issues for Congress* by Ronald O'Rourke. Washington, DC, 2013, 11-12.

[47] Heather Conley, Terry Toland, and Jamie Kraut. *A New Security Architecture for the Arctic: An American Perspective* (Washington, DC: CSIS, January 2012), 11.

[48] Ibid.,11.

UNCLOS, this could place the US at the center of an international quandary.

As the Russian Federation projects their power in the Arctic via eighteen icebreakers, six of which are nuclear, the US finds itself extremely limited in its ability to project its power via surface ships in the Arctic.[49] The US Navy (USN) currently has no surface ships capable of operating in the icy Arctic waters while the USCG has only three polar icebreakers.[50] There are two heavy icebreakers, USCGC POLAR STAR (WAGB-10), and USCGC POLAR SEA (WAGB-11), capable of breaking ice six feet thick at three knots (continuous) and twenty-one feet backing and ramming and one medium icebreaker, USCGC HEALY (WAGB-20), capable of breaking ice four and a half feet at three knots (continuous) and eight feet backing and ramming.[51]

---

[49] Scott G. Borgerson, "Arctic Meltdown: The Economic and Security Implications of Global Warming," *Foreign Affairs*, vol 87, no. 2. (March/April 2008): 63-77; and Eve Conant, "Breaking the Ice: Russian Nuclear-Powered Ice-Breakers," Scientific American Guest Blog, entry posted September 8, 2012, http://blogs.scientificamerican.com/guest-blog/2012/09/08/breaking-the-ice/ (accessed January 5, 2013).

[50] U.S. Department of Defense, *Report to Congress on Arctic Operation and the Northwest Passage*. (Washington, DC, May 2011), 17.

[51] U.S. Coast Guard,"USCGC POLAR SEA (WAGB 11)," http://www.uscg.mil/pacarea/cgcpolarsea/; and U.S. Coast Guard "CGC Healy Ship's Characteristics," http://www.uscg.mil/pacarea/ cgchealy/ship.asp (accessed Feb 9, 2013).

Figure 2. CGC POLAR SEA and CGC POLAR STAR break ice in the Antarctic in 1995.

Source: USCG photo by ETCS Wayne Jarvis

Currently, the POLAR STAR is in drydock undergoing major repairs, POLAR SEA is in a caretaker status, thus leaving HEALY as the only US polar icebreaker in service.[52]

USG NATIONAL POLICY

On January 9, 2009, President George W. Bush signed National Security Presidential Directive 66 - Homeland Security Presidential Directive 25 (NSPD 66/HSPD 25), giving the US a national policy on the Arctic and establishing six overarching elements of policy: national security, environmental protection, management of natural resources, increased international cooperation, partnering with indigenous Arctic people, and enhanced scientific research. To accomplish these six goals, the policy further breaks down into seven categories and assigns implementation responsibilities to various departments. These seven categories make up the basis for evaluating US implementation of policy.

The first category of National Security and Homeland Security discusses the US rights as a sovereign nation to provide missile defense and early warning, strategic deterrence, airlift, maritime security operations to include law enforcement, maritime presence and freedom of navigation and overflight. Additionally it affirms that the Northwest Passage is an international strait and states that parts of the Northern Sea Route includes international straits. It declares that the US must assert a national presence in the Arctic, specifically calling "to project sea power" in the Arctic and maintain jurisdiction over the US EEZ and continental shelf. It also addresses concerns over border security and terrorist activities, calling them "fundamental homeland security interests." Under International Governance the policy focuses mainly on the Arctic Council, but also calls for the US to cooperate with numerous nations and organizations including the creation of new ones if necessary. It also asserts that the US Congress

---

[52] This USCG does have a large icebreaker, USCGC MACKINAW (WLBB-30), solely dedicated to the Great Lakes and smaller icebreakers working inland rivers and near coastal areas.

should ratify UNCLOS and that an Arctic treaty is not appropriate. With regard to the Extended Continental Shelf and Boundary Issues, the policy restates that the most effective way for the US to make a continental shelf claim is via UNCLOS. It also acknowledges the boundary disputes with Canada and the Russian Federation. In the area of Promoting International Scientific Cooperation, the policy recognizes that scientific research is critical, and encourages sharing of Arctic research platforms and information. Regarding Maritime Transportation, the policy aims to provide safe and reliable navigation while protecting maritime commerce and the environment. The policy accepts that Economic Issues, including Energy, poses numerous challenges. However, the policy affirms that the US must continue to work with international organizations, including regulatory ones in order to maximize the economic potential of the Arctic. In the area of Environmental Protection and Conservation of Natural Resources, the policy determines that based on the limited data and understanding of the Arctic, efforts in the Arctic should be risk-based. The policy also admits that the Arctic is unique and that steps must be taken to ensure that decisions are made based on accurate information and that the environment must be protected. It specially states that the US supports international agreements on fisheries management and pollution control.[53]

US DEPARTMENTS AND AGENCIES POLICIES

NSPD 66/HSPD 25 is only one element of US National Policy, but the one that all department and agency policies should be based. The Department of Defense policies are numerous and found in various papers and reports. The US Navy (USN) was quick to respond to national guidance in May 2009 by assigning Task Force Climate Change (TFCC). TFCC developed the US Navy's Arctic Roadmap, which the Navy published in October 2009. This roadmap laid out the Navy's plan to implement NSPD 66/HSPD 25 over five years using a three-phase process. The roadmap laid out six objectives each with

---

[53] President, *National Security Presidential Directive/NSPD 66 – Homeland Security Presidential Directive /HSPD 25* (January 9, 2009).

several action items. These objectives focused on more planning and assessments. Not until phase three and the 2014 budget cycle does the Navy state that it will begin applying funding to Arctic requirements.[54] The Navy followed with its "first deliverable" under the Arctic Roadmap on May 21, 2010 by releasing a four-page document on the Navy Strategic Objectives in the Arctic Region which states, "[t]he Navy's desired end state is a safe, stable and secure arctic region where U.S. national and maritime interests are safeguarded and the homeland is protected."[55]

Also in 2010 the Navy along with the US Marines (USMC) and the US Coast Guard (USCG) released the Naval Operations Concept. This document recognizes national security interests in the Arctic. It states that the three naval services are ready to operate in the Arctic, but continues to list numerous broad challenges associated with operating in the Arctic, which include, "lack of environmental awareness, navigation capabilities, and supporting infrastructure, as well as competing jurisdictional and resource claims." [56] The document goes on to further state that in order to fulfill US national interests that submarines and icebreakers or ice capable ships will have to deploy routinely to the Arctic. It also reinforces that icebreakers are critical to enforcing US Sovereignty in the Arctic and that "they are the only means of providing assured surface access in support of Arctic maritime security and sea control missions." [57]

On March 30, 2011, shortly before President Obama signed a change to the Unified Command Plan (UCP) assigning US Northern Command as the sole Geographic Combatant Commander responsible for planning in the Arctic region, the Commander of US Northern Command (NORTHCOM), Admiral James

---

[54] U.S. Department of the Navy, *U.S. Navy Arctic Roadmap* (Washington, DC, October 2009).

[55] U.S. Navy, Chief of Naval Operations memo dated 21 May 2010, *Navy Strategic Objectives for the Arctic* (Washington, DC, 2010), 1 and 4.

[56] U.S. Navy, U.S. Marine Corps, U.S. Coast Guard, *Naval Operations Concept 2010: Implementing the Maritime Strategy* (Washington, DC, 2010), 32.

[57] Ibid., 91.

Winnefeld, Jr, presented NORTHCOM's posture statement before the House Armed Services Committee.[58] In his posture statement, ADM Winnefeld states that although he has confidence in the US Missile Defense capabilities, he would like to have improved sensors, weapons systems, and better-trained operators. He goes on to state that he has made the Arctic a "key focus" area, but falls short of providing any real details. He makes three broad statements concerning NORTHCOM's actions to include; constructing a Commander's Intent, fostering a better relationship with Canada Command, and "maturing our understanding of our gaps" in the Arctic. He continues by stating the gaps are: domain awareness, communications, shore infrastructure, icebreaking, search and rescue (SAR), ocean charting and Arctic change forecasting.[59]

Then in May 2011, in response to a Congressional request, DoD provided a report on the Arctic and the Northwest Passage. In this report, DoD presented an overview, largely taken from the above documentation, of DoD's actions with regard to the Arctic. On page one, it reaffirms the Navy's assessment of what the President said in NSPD 66/HSPD 25; "The overarching strategic national security objective is a stable and secure region where U.S. national interests are safeguarded and the US homeland is protected." [60] It also points out the same challenges as NORTHCOM's posture statement, except DoD points out that SAR is not a DoD primary mission and belongs to the USCG. It focuses on DoD not being "late-to-need" with regard to Arctic capabilities and warns about being too early. The report also states that the US needs "assured Arctic access"[61] which is the same wording used in the Naval Operations

---

[58] Congressional Research Service, *Changes in the Arctic: Background and Issues for Congress* by Ronald O'Rourke. Washington, DC, 2013, 48.

[59] U.S. Congress. House Committee on Appropriations. *Statement of Admiral Jams A. Winnefeld, Jr, U.S. Navy, Commander U.S. Northern Command and North American Aerospace Defense Command: Hearing before the House Armed Services Committee.* 112 Cong. 2nd sess., March 6, 2012.

[60] U.S. Department of Defense, *Report to Congress on Arctic Operation and the Northwest Passage.* (Washington, DC, May 2011), 1.

[61] Ibid., 3.

Concept that stated that USCG icebreakers are the only surface means that the US has for "assured surface access." [62] DoD also continually stresses that the rate of change in the Arctic is almost impossible to determine and that the US should study the Arctic changes for a longer period before spending money on Arctic capable assets. [63] However, in order to completely and competency study the Arctic the US needs more icebreakers. The report attempts to follow NSPD 66/HSPD 25 by listing the six overarching elements of policy, but never directly engages in the five specific concerns listed in the National/Homeland Security section nor the five-implementation steps under that section. The report repeatedly stresses that the threat assessment for National Security in the Arctic is low and, therefore, DoD is capable of meeting current requirements. However, it goes on to give details on three gaps in capabilities. Specifically, it points out communication difficulties in High-Frequency (HF) radios above 70 degrees North, Global Positioning System (GPS) performance in the Arctic and lack of ice characterization charts. [64]

Service specific capabilities are also included in the report, and it also acknowledges that the USN has no surface ships that are capable of operating in or near ice. The US Army is capable of conducting missile defense, training and exercises in the Arctic. The Army has two Brigade Combat Teams and a National Guard Infantry Brigade and aviation unit in Alaska. The U.S. Army Corps of Engineers has a Cold Regions Research and Engineering Lab (CRREL) in New Hampshire that develops solutions to military challenges in the Earth's cold regions. The report then focuses on units assigned within the Arctic, but provides no assessment of their capability. The US Air Force's (USAF) Arctic capabilities center on its Ballistic Missile Early Warning System, its Reserve units with LC-130 Hercules that are ski-

---

[62] U.S. Navy, U.S. Marine Corps, U.S. Coast Guard, *Naval Operations Concept 2010: Implementing the Maritime Strategy* (Washington, DC, 2010), 91.

[63] U.S. Department of Defense, *Report to Congress on Arctic Operation and the Northwest Passage.* (Washington, DC, May 2011), 3.

[64] Ibid., 3, 15-16.

equipped and on its ownership of Thule Air Base, which provides a deep water port and air mobility. The USAF also provides an Enhanced Polar System (EPS) to support Extremely High-Frequency satellite communications. The report states that the USMC is ready to respond in any climate and any place, but other than listing cold weather training, provides not details regarding capability.[65] The report does not include the USCG except for its capabilities to support National Defense Missions and specifically addresses the lack of icebreakers, but then fails to address fully the need for icebreakers, even though Congress gave specific tasking regarding a need assessment for additional icebreakers.[66] The report also discusses shore infrastructure, but again ascertains that the unpredictability of the Arctic rate of change preempts the need for an immediate increase in shore infrastructure except maybe airport and hanger facilities for the USCG.[67]

In August 2011, the Navy released the *Arctic Environmental Assessment and Outlook Report*, which supports the Navy's Arctic Roadmap. This report provides scientific data that supports DoD's assessment of the difficulty in predicting the rate of change in the Arctic. However, it clearly states that at some time in the future there will be an ice-free Arctic during the summer months. It also gives specific details to challenges the Navy would face while operating in the Arctic. It goes onto say that it is unlikely that the Navy will shift its current deployment schedules to increase presence in the Arctic. It does however, specifically point out that the USCG is "expected to maintain" its icebreaking fleet and that the Navy is expected to be ready to operate in the Arctic.[68]

---

[65] Ibid., 17-19.

[66] Ibid., 29-32. and President, National Security Presidential Directive/NSPD 66 – Homeland Security Presidential Directive /HSPD 25 (January 9, 2009); and U.S. Government Accountability Office, Report to Congressional Committees, *Arctic Capabilities: DOD Addressed Many Specified Reporting Elements in Its 2011 "Arctic Report" but Should Take Steps to Meet Near-and-Long-term Needs*, GAO - 12-180(Washington, DC, January 2012), 10.

[67] U.S. Department of Defense, *Report to Congress on Arctic Operation and the Northwest Passage.* (Washington, DC, May 2011), 25.

[68] U.S. Navy, *Arctic Environmental Assessment and Outlook Report; In support of the Navy*

Focusing on DHS, documentation reveals that the USCG has been discussing the need for replacement and/or additional icebreakers even before the current US Policy. The National Research Council filed a report in 2007 that assessed the need for USCG icebreakers to conduct research in the Arctic and Antarctic. The document points out the underfunding and lack of maintenance on the icebreakers and reaffirms the national need for polar icebreakers calling them, "essential instruments of US national policy." [69] In February 2008, the Congressional Research Service provided its first report to congress regarding USCG polar icebreaker modernization. The documented pointed out that the two existing polar class heavy icebreakers were beyond their thirty-year service life and either needed to be overhauled or replaced. [70] In 2010, the USCG conducted what is called the High Latitude Study (HLS), which provided an analysis of USCG requirements and capability gaps. The study pointed out that the USCG, as a multi-mission service, is closely linked to DoD and presently holds the nation's only icebreakers. It highlights that the USCG's primary gap in Arctic operations is the gap in polar icebreaker capability as both the USCGC POLAR SEA and USCGC POLAR STAR face enormous maintenance issues as well as lack of sea time for the crew due to numerous loss of operational days because of maintenance. [71] The HLS also indicated gaps in communications and shore infrastructure. Most importantly, the study revealed that in order for the USCG to fulfill all its non-national defense missions, in the Arctic and Antarctic, it needs three heavy and three medium icebreakers. To fulfill its national defense mission as laid out by the Naval Operations Concept, it needs now an additional three heavy and

---

Arctic Roadmap; Action Item 5.7 (Washington, DC, 2011), 22.

[69] National Research Council, *Polar Icebreakers in a Changing World: An Assessment of U.S. Needs*, National Acadmey of Sciences (Washington, DC, 2006), 1-4.

[70] Congressional Research Service, *Coast Guard, Polar Icebreaker Modernization Background, Issues, and Options for Congress* by Ronald O'Rourke. Washington, DC, 2008, Non-numbered summary page.

[71] ABS Consulting, *United States Coast Guard High Latitude Region Mission Analysis Capstone Summary* (Arlington, VA: ABS Consulting), 10.

one medium icebreakers.[72] However, in February 2010 DHS published its first Quadrennial Homeland Security Review (QHSR) and in February 2012 released the DHS Strategic Plan (DHSSP). The QHSR and the DHSSP are DHS's first attempt at linking strategy to performance, but makes no mention of the Arctic.[73] However, USCG officials published its guidance in COMDTINST 16003.1 *US Coast Guard Arctic Strategic Approach*, which lays out mission execution on the Arctic and calls for increased partnerships, domain awareness, use of technology advances and "right-sized assets." The instruction also promised to work to get the right force structure and equipment to carry out the USCG's Arctic missions.[74] Without outwardly saying it, the instruction alluded to the fact that the USCG needs additional icebreakers.

To increase its Arctic domain awareness, the USCG began making seasonal biweekly over-flights of the Arctic in October of 2007. During this first season of over-flights, the USCG reported the challenges of communications and lack of shore infrastructure for the C130 airframe.[75] This began as gradual increase in USCG summer presence in the Arctic culminating in Arctic Shield 2012 in which the USCG deployed to the Arctic the 418ft high endurance cutter USCGC BERTHOLF, the 282 ft medium endurance cutter USCGC ALEX HALEY and two sea-going light ice-capable buoy tenders, the USCGC HICKORY and USCGC SYCAMORE. They also rented a forward operating base in Barrow, deploying two MH-60 helicopters along with air, ground, and communication crews. Also during the summer, the USCG conducted exercises with NORTHCOM, the USN, and exercised its own oil spill skimming

---

[72] Ibid., 2.

[73] U.S. Department of Homeland Security, *Quadrennial Homeland Security Review Report,* (Washington, DC, February 2010); and U.S. Department of Homeland Security, *Strategic Plan Fiscal Years 2012-2016,* (Washington, DC, February 2012).

[74] U.S. Coast Guard, Commandant Instruction 16003.1, *U.S. Coast Guard Arctic Strategic Approach*, (Washington, DC, April 26, 2011), 1-4.

[75] U.S. Government Accountability Office, Report to Congressional Committees, *Coast Guard, Efforts to Identify Arctic Requirements are ongoing, but More Communication about Agency Planning Efforts Would be Beneficial*," GAO -10-870, 28.

equipment, which was carried aboard the sea-going buoy tenders. At the end of the summer, the Arctic crews were responsible for saving or assisting ten people while increasing Arctic domain awareness.[76]

Figure 3. CGC WILLOW, sister ship to CGC HICKORY and CGC SYCAMORE transits by an iceberg on an Arctic patrol.

Source: U.S. Coast Guard photo by Petty Officer 3rd Class Luke Clayton.

Figure 4. USCGC BERTHOLF in Arctic Ocean, Sept. 14, 2012 as part of Arctic Shield 2012.

---

[76] U.S. Coast Guard, Coast Guard District 17 External Affairs Office news release, "Imagery Available: Coast Guard completes Arctic Shield 2012," 1 Nov 2012, http://www.uscgnews.com /go/doc/4007/1594651/.

The Department of State's Office of Ocean and Polar Affairs (OPA) provides its overarching goal through four main themes: ratification of UNCLOS, bilateral and multipateral polar agreements, active leadership in international Arctic groups and coordination with other US departments and agencies. The OPA policy also provide ten very specific objectives that include: ratification of UNCLOS, environmental protection, conserve marine life, improve security including "protect freedom of navigation", promote peace, engagement in Arctic Council and other international groups, promote scientific research, establish US extended continental shelf, and protect cultural heritage.[77]Additionally in 2011, then Secretary of State, Hillary Clinton personally led the US delegation at the Arctic Council meeting in Greenland setting the standard for the department.[78]

In February 2011, the Department of Commerce published *NOAA's Arctic Vision & Strategy,* which provides the agency's vision:

> NOAA envisions an Arctic where decisions and actions related to conservation, management, and use are based on sound science and support healthy, productive, and resilient communities and ecosystems. The agency seeks a future where the global implications of Arctic change are better understood and predicted.[79]

NOAA's report goes on to provide six focus areas for the Arctic: sea ice forecasting, foundational science, weather forecasts and warnings, international and national partnerships, ocean management, and Arctic communities.[80]

---

[77] U.S. Department of State, "Ocean and Polar Affairs," http://www.state.gov/e/oes/ocns/opa/.

[78] Congressional Research Service, *Changes in the Arctic: Background and Issues for Congress* by Ronald O'Rourke. Washington, DC, 2013, 41.

[79] U.S. National Oceanic and Atmospheric Administration, *NOAA's Arctic Vision & Strategy* (Washington, DC, February 2011), 6.

[80] Ibid., 7.

IMPLEMENTATION OF NSPD 66 – HSPD 25

National Security

In NSPD 66/HSPD 25, the President stated the nation's number one priority was to "meet national security and homeland security needs relevant to the Arctic region."[81] The policy specifically tasks DoS, DoD, and DHS with five specific implementation steps. The Department of State continues active participation in the Arctic Council. However, the Arctic Council's charter specifically states that it does not engage in military matters.[82] Nonetheless, the US's active engagement in the Arctic Council has a large impact on National and Homeland Security and is discussed in depth in a later section. However, DoD and DHS have large rolls in this arena. In addressing the first implementation step of the USG policy, "[d]evelop greater capabilities and capacity, as necessary, to protect United States air, land, and sea borders in the Arctic region," DoD appears to have not developed any additional capabilities.[83] Yet, most of DoD's documents express limits of surface ships operating in the Arctic and points out that submarines often operate under the surface of the Arctic. The US Navy does maintain an Arctic Submarine Laboratory to provide expertise and "assured access" to the Arctic with its submarine fleet.[84] The national policy also says "as necessary" and DoD has stated numerous times that based on risk it sees no need to increase Arctic capabilities with the exception of USCG icebreakers and maybe some shore infrastructure for the USCG. DHS too has failed in significantly increasing its Arctic capabilities, with the sole exception of the USCG's summer presence during Arctic Shield 2012. Using risk analysis, this might be a reasonable method for the short term; the long-term solution of increased icebreaking capability is

---

[81] President, *National Security Presidential Directive/NSPD 66 – Homeland Security Presidential Directive /HSPD 25* (January 9, 2009).

[82] Arctic Council, *Declaration on the Establishment of the Arctic Council* (Ottawa, 1996), 1.

[83] President, *National Security Presidential Directive/NSPD 66 – Homeland Security Presidential Directive /HSPD 25* (January 9, 2009).

[84] Submarine Force U.S. Pacific Fleet, "Arctic Submarine Laboratory," http://www.csp.navy.mil/ asl/index.htm (accessed February 5, 2013).

still ongoing with no sure solution on the horizon.

In the second step, "[i]ncrease Arctic maritime domain awareness in order to protect maritime commerce, critical infrastructure, and key resources," again DoD has completed a number of assessments, formed a working team with DHS, and stated that it has a challenges with domain awareness, but, has done little to improve its maritime domain awareness. [85] DHS has also done little outside of the USCG's Arctic Shield. Certainly, the USCG has increased its understanding of the Arctic through its experiences during summers; however, its ability to know what is going on the Arctic with regard to National Security and true Arctic domain awareness remains limited.

In the third step, "[p]reserve the global mobility of United States military and civilian vessels and aircraft throughout the Arctic region" DoD has maintained status quo and continues to only be able to operate in the Arctic with submarines, and relies solely on the USCG for surface assets. [86] With regard to civilian vessels, the US is largely reliant upon foreign assistance as was the case in July 2007 when a Canadian Coast Guard Cutter, along with a private helicopter, assisted a disabled US vessel. In addition, in January 2012 when the town of Nome Alaska needed emergency fuel, the town could only find one willing Russian double-hull, ice-class tanker to deliver the necessary fuel. The USG had to give the Russian ship a waiver to the Merchant Marine Act of 1920, allowing it to deliver fuel from one US port to another US port. [87] Even then, successful delivery was dependent upon the USCG's only operational icebreaker leading the way. With the USCG's heavy icebreakers both non-operational at the time, a US

---

[85] President, *National Security Presidential Directive/NSPD 66 – Homeland Security Presidential Directive /HSPD 25* (January 9, 2009); and U.S. Government Accountability Office, Report to Congressional Committees, *Arctic Capabilities: DOD Addressed Many Specified Reporting Elements in Its 2011 "Arctic Report" but Should Take Steps to Meet Near-and-Long-term Needs*, GAO -12-180, 13.

[86] Nicole Klauss, "US Navy lacks ability to operate in Arctic, games reveal," *Anchorage Daily News*, April 28, 2012, http://www.adn.com/2012/ 04/28/2444408/us-navy-lacks-ability-to-operate.html (accessed February 4, 2013).

[87] The Merchant Marine Act of 1920 states that a foreign-flagged vessel may not transport goods from one US port to another US port.

city was completely dependent on the USCG's medium icebreaker, USCGC HEALY.[88] A few more feet of ice thickness and the US would have been forced to call for foreign assistance, most likely the Canadian Coast Guard, to provide the requisite heavy icebreaker capability.

Figure 5. USCGC HEALY breaks ice for the 370-foot Russian tanker RENDA on January 14, 2012. *Source:* U.S. Coast Guard photo by Petty Officer 2nd Class Charly Hengen.

In the fourth step, "[p]roject a sovereign United States maritime presence in the Arctic in support of essential United States interest," DoD has maintained the status quo in this arena. DoD has extremely limited ability to project power in the maritime domain, yet, ascertains that the USCG via its icebreakers are the primary assets the US has to project highly visible presence. The USCG proved that it could projected limited US sovereignty during the summer months, but sorely lacks that ability during the majority of the year with only one functional icebreaker.

---

[88] Mia Bennett, "The Icebreaker That Could: USCGC HEALY Leads Way for Russian Tanker," *Foreign Policy Association Blog*, entry posted January 10, 2012, http://foreignpolicyblogs.com /2012/01/10/the-icebreaker-that-could-uscgc-healy-slowly-leads-the-way-for-russian-ship-renda-carrying-fuel-to-nome/ (accessed February 4, 2013).

The fifth implementation step is to, "[e]ncourage the peaceful resolution of [boundary] disputes in the Arctic region." DoS's US – Canada Relations Fact Sheet of June 29, 2012 shows tremendous efforts between the US and Canada to cooperate in the Beyond the Border initiative and other law enforcement agreements to ensure security across their borders and improve legitimate travel between the countries, however, there is nothing regarding a resolution of boundary disputes.[89]

Overall, it is clear that the US lacks the capability to operate effectively in the Arctic if a national /homeland security emergency were to occur. DoD continues to rely on the USCG to provide surface assist despite the fact that the USG policy, and their own documents assert that it is the US Navy's job to ensure freedom of the sea.[90] Several 'what if' scenarios exist, such as pirates attacking a US flagged vessel carrying US military equipment or the catastrophic accident involving a US submarine in which the US has extremely limited capability to respond. As Professor Walter Berbrick, of the Center for Naval Warfare Studies points out, "We have limited capability to sustain long-term operations in the Arctic due to inadequate icebreaking capability. The Navy (and the country) finds itself entering a new realm as it relates to having to rely on other nations."[91]

International Governance

NSPD 66/HSPD 25 also gives the President's direction for implementing international governance in the Arctic, by listing four specific steps.[92] The USG policy charges the Department of State

---

[89] The Beyond the Border initiative is a joint endeavor between the US and Canada to enhance the security of their borders, while facilitating commerce across these borders; U.S. Department of State, "U.S. Relations with Canada," http://www.state.gov/r/pa/ei/bgn/2089.htm (accessed March 14, 2013).

[90] U.S. Department of Defense, *Report to Congress on Arctic Operation and the Northwest Passage.* (Washington, DC, May 2011), 3-4.

[91] Words in parenthesis added by author for effect; and Nicole Klauss, "US Navy lacks ability to operate in Arctic, games reveal," *Anchorage Daily News*, April 28, 2012, http://www.adn.com/2012/04/28/2444408/us-navy-lacks-ability-to-operate.html (accessed February 4, 2013).

[92] President, *National Security Presidential Directive/NSPD 66 – Homeland Security Presidential Directive /HSPD 25* (January 9, 2009).

to work through the United Nations and other international frameworks and to make appropriate recommendations to the Arctic Council. DoS serves as the primary led on the Arctic Council, however, the USCG has increasingly become the US representative on many multilateral coordinating groups. A Coast Guard Officer normally serves as the head of delegation to the IMO committee and sub-committee meetings and serves as the primary US representative for development of IMO policy.[93] December 9-11, 2009, the Department of State and USCG hosted the first Arctic SAR Task Force meeting in Washington, D.C., with the USCG being the US lead.[94] Also, the Department of State via its Office of Ocean and Polar Affairs continues to work within the Arctic Council and it working groups to promote US interests in the Arctic.[95] Moreover, the US played a pivotal role as one of three nations who took the lead in the Arctic Marine Shipping Assessment (AMSA) report.[96]

The USG policy encourages departments and agencies to "[c]ontinue to seek advice and consent of the United States Senate to accede to the 1982 Law of the Sea Convention." DoD, DoS, DoC, and DHS have all publically stated the need for Congress to ratify UNCLOS. The US Navy, however, gives some pushback in its Arctic Roadmap by stating that the US Navy will support US accession of UNCLOS as "applicable to Navy's interests," which are listed as freedom of navigation, treaty vs. customary law, environmental laws, and extended continental shelf claims. The Navy Roadmap includes a statement that it would develop "talking points, information papers, or briefings for senior Navy leadership and Congressional staffs as requested."[97] Next, the Navy goes on to give what on the surface is a mundane

---

[93] Jon Trent Warner, LCDR, USCG, "Supporting the Arctic Marine Transportation System," *USCG Proceedings* (Summer 2011), 68.

[94] U.S. Department of State, "Arctic Search and Rescue," http://www.state.gov/e/oes/ocns/opa/arc/c29382.htm (accessed March 14, 2013).

[95] Jon Trent Warner, LCDR, USCG, "Supporting the Arctic Marine Transportation System," *USCG Proceedings* (Summer 2011), 68.

[96] Arctic Council, *Arctic Marine Shipping Assessment 2009 Report*, April 2009.

[97] U.S. Department of the Navy, *U.S. Navy Arctic Roadmap* (Washington, DC, October 2009). 11.

politically correct statement by saying that it will continue to emphasis its public statement "that the Navy is committed to being responsible stewards of the environment. While being committed to conducting military readiness activities in an environmentally sound manner, the Navy is opposed to any framework which unreasonably restricts or prevents our ability to train and operate effectively." [98] The statement clearly gives pushback to working within international organizational frameworks.

On July 27, 2011, in a testimony before the Subcommittee on Oceans, Atmosphere, Fisheries, and Coast Guard of the Committee on Commerce, Science and Transportation, the Commandant of the Coast Guard, Admiral Robert Papp stated,

> As a matter of policy and stewardship, we encourage the Senate to ratify the Law of the Sea Treaty. Law of the Sea has become the framework for governance in the Arctic. Every Arctic Nation except the United States is a party. As our responsibilities continue to increase in direct proportion to the Arctic's emerging waters, it is more vital than ever that the U.S. ratified to Law of the Sea. [99]

Regardless of all the rhetoric regarding UNCLOS, the policy is truly about working with other nations and fostering relationships for multilateral agreements. Overall, the US has been aggressive in participation with the Arctic Council and working with multinational groups. But, there has been little discussions regarding working with non-Arctic nations. China's Rear Admiral Yin Zhuo has made it clear that other nations feel that the Arctic does not belong to just the Arctic nations. "The Arctic belongs to all the people around the world as no nation has sovereignty over it. China must play an indispensable role in Arctic exploration as we have one-fifth of the world's population." [100]

---

[98] Ibid.,12.

[99] U.S. Congress. Senate Committee on Commerce, Science, and Transportation. *Testimony of Admiral Robert Papp Commandant, U.S. Coast Guard on "Arctic Operations": Hearing before the Subcommittee on Oceans, Atmosphere, Fisheries, and Coast Guard.* 112 Cong. 1st sess., July 27, 2011.

[100] David Akin, "Harper deals with new Arctic rival: China", *Toronto Sun*, June 23, 2010. http://www.torontosun.com/news/g20/2010/06/22/14484401.html (accesses Mar 6, 2013).

Extended Continental Shelf and Boundary Issues

Few Americans know that the US has active boundary disputes with Canada and the Russian Federation. In addition, even fewer Americans know that the US has the potential to make claims for an extended Continental Shelf in the Arctic, which could hold vast amounts of natural gas and oil. NSPD 66/HSPD 25 directs the State Department and other responsible agencies to address these issues via three policy implementation steps. In regards to the first, "[t]ake all actions necessary to establish the outer limit of the continental shelf appertaining to the United States, in the Arctic and in other regions, to the fullest extent permitted under international law," the State Department, DHS, and DoC have all taken action in anticipation of Senate consent of UNCLOS. After accepting UNCLOS, nations have ten years to file an extended Continental Shelf claim. In order to substantiate a claim, the US formed a continental shelf Task Force that includes the State Department, National Oceanic and Atmospheric Administration (NOAA) and the USCG. NOAA created the official US nautical charts, which show the limits of the US territorial sea, contiguous zone, and the Exclusive Economic Zone (EEZ) in the Arctic as previously declared by the US and in accordance with UNCLOS.[101] NOAA continues to chart US baseline and EEZ claims as scientific data changes. The US and Canadian Coast Guards joined forces in a Hydrographic Commission in an effort to map the North American extended continental shelf. In 2009 and 2010, USCGC HEALY and CCGS LOUIS S. ST-LAURENT worked together with scientists onboard as part of US Department of State sponsored Extended Continental Shelf Project. USCGC HEALY collected data on the shape of the seafloor while CCGS LOUIS S. ST-LAURENT collected data on sediment thickness and layers beneath the seafloor.[102] CCGS LOUIS S. ST-LAURENT can collect seismic reflection data, which the US needs, while USCGC HEALY cannot; while USCGC HEALY is better equipped to collect

---

[101] Megan L. Campbell, "United States Arctic Ocean Management & the Law of the Sea Convention," (externship paper, U.S. Department of Commerce, undated), 2.

[102] Extended Continental Shelf Project, "2010 Extended Continental Shelf Survey," http://continentalshelf.gov/missions/10arctic/welcome.html (accessed March 17, 2013).

multi-beam bathymetric data, which Canada needs.[103] Therefore, the joint endeavor appears to the best method for both countries to obtain scientific data for extended continental shelf claims.[104]

The second implementation element is to "consider the conservation and management of natural resources during the process of delimiting the extended continental shelf." There appears to be no specific study or activity within the US government to address this specific issue. However, scientists and crewmembers aboard USCGC HEALY follow national and international laws regarding the environment while collecting data in the Arctic.

Figure 6. USCGC HEALY and CCGS LOUIS S. ST-LAURENT in the Arctic Ocean Sept. 5, 2009.
Source: U.S. Coast Guard photo by Petty Officer Patrick Kelley.

---

[103] There are two types of data needed to determine the Continental Shelf limits. One is bathymetric data that provides a three-dimensional map of the ocean floor. The second is seismic reflection data, which essentially provides the thickness, geometry, and other characteristics of the ocean floor as described at http://www.state.gov/e/oes/continentalshelf/.

[104] U.S. Department of State, "Defining the Limits of the U.S. Continental Shelf," http://www.state.gov/e/oes/continentalshelf/ (accessed March 19, 2013).

The third element is to "continue to urge the Russian Federation to ratify the 1990 United States-Russia maritime boundary agreement." The US, through the Department of State, maintains a diplomatic relationship with Russia via a US Ambassador in Moscow and likewise a Russian Ambassador serves in Washington. While the State Department boasts many improvements regarding relations with Russia, such as adoption and visas, the US-Russian maritime boundary does not appear to be a topic on either nation's agenda. In fact, soon after the issuance of NSPD 66/HSPD 25, the new president, Barack Obama, created the U.S.-Russia Bilateral Presidential Commission which has nineteen working groups none of which focus on the maritime boundary.[105]

In review, the US has established at least three bodies (a task force and two bi-lateral commissions) that will no doubt assist in resolving the continental shelf and boundary issues in the Arctic; however, they fall far short in implementing the specific guidance given in NSPD 66/HSPD 25. It is also clear that the US must conduct more research in order to make an extended continental shelf claim in the Arctic. Again, the national capability of a US icebreaker is proving critical to US national policy in the Arctic.

<u>Promoting International Scientific Cooperation</u>

NDPD 66 – HSPD 25 also provides a detailed policy of promoting international scientific cooperation and specifically addresses several areas. As documented below this area of the USG policy is one of the most well implemented segments to date via grants to universities and individual scientists. The National Science Foundation and NOAA have excelled in all areas of policy implementation. However, the individual scientists and American universities deserve most of the credit.

---

[105] U.S. Department of State, U.S.- Russia Bilateral Commission. http://www.state.gov/p/eur/ci/rs/usrussiabilat/index.htm (accessed March 19, 2013).

NSPD 66 – HSPD 25 emphasizes six critical parts to policy implementation as it relates to promoting scientific international cooperation.

The first of these is to "[c]ontinue to play a leadership role in research throughout the Arctic region." The US has made efforts in this endeavor mainly through the National Science Foundation (NSF), who has the lead role for the US Arctic research.[106] The NSF chairs the Interagency Arctic Research Policy Committee (IARPC) and through its Division of Arctic Sciences coordinates US Science observations in the Arctic.[107] Most recently, IARPC published its five-year (FY13 – FY17) plan to coordinate federally funded research projects.[108] There is also a US Arctic Research Committee (USARC) comprising of eight members, seven appointed by the President and the eighth coming from the NSF. The USARC's primary purpose is to recommend national Arctic research policy to the President and Congress, additionally; they provide a daily Arctic Update via e-mail.[109] In February 2013, they published a *Report on the Goals and Objectives for Arctic Research 2013–2014*, which provided five research goals to Congress. The goals included: observe, understand, and respond to environmental change in the Arctic; improve Arctic human health; understand natural resources; advance civil infrastructure research; and assess indigenous languages, identities, and cultures.[110]

USG Policy also calls to "[a]ctively promote full and appropriate access by scientists to Arctic

---

[106] National Science Foundation, Fact Sheet, "The Arctic Observing Network (AON)" July 10, 2007. http://www.nsf.gov/news/news_summ.jsp?cntn_id=109687 (accessed March 19, 2013).

[107] Ibid.

[108] National Science Foundation, "Interagency Arctic Research Policy Committee (IARPC) – Arctic Research Plan," http://www.nsf.gov/od/opp/arctic/iarpc/arc_res_plan_index.jsp (accessed March 19, 2013).

[109] U.S. Arctic Research Commission, "About USARC," http://www.arctic.gov/about.html (accessed March 19, 2013).

[110] U.S. Arctic Research Commission. *Report on the Goals and Objectives for Arctic Research 2013–2014,* (Arlington, VA. February 2013), 1-3.

research sites through bilateral and multilateral measures and by other means." [111] As previously

discussed, the US has teamed up with the Canadian Government to conduct Arctic research. [112]

Additionally, the NSF and NOAA work hand in hand with Finnish and Russian Science Teams at the

Tiksi International Hydro meteorological Observatory in Tiksi, Russia. [113] One of the most substantial

international Arctic research provisions is that of US scientists working aboard Russian Federation

Research vessels. In 2003, NOAA signed a Memorandum of Understanding (MOU), with the Russian

Federation, that enabled sharing of research data across government agencies and educational institutions.

The MOU encouraged joint research projects and created a Joint Coordinating Committee. [114] This MOU

has fostered a series of joint scientific cruises aboard Russian vessels. The most recent of these cruises

was aboard the Russian research vessel PROFESSOR KHROMOV as part of the 2012 Russian-American

Long-term Census of the Arctic (RUSALCA) cruise in which scientists documented marine mammals

along the Chukotka coast of Russia. [115]

Through the Sustaining Arctic Observing Networks (SAON), the US continues to "[l]ead the

effort to establish an effective Arctic circumpolar observing network with broad partnership from other

relevant nations" a required by NSPD 66 – HSPD 25. SAON is the international observing network body

---

[111] President, *National Security Presidential Directive/NSPD 66 – Homeland Security Presidential Directive /HSPD 25* (January 9, 2009).

[112] Extended Continental Shelf Project, "Extended Continental Shelf Summary of Missions," http://continentalshelf.gov/missions.html (accessed February 3, 2013).

[113] International Arctic Systems for Observing the Atmosphere, Tiksi, Russia, "Tisksi International Hydrometeorological Observatory," http://iasoa.org/iasoa/index.php?option= com_content& task=view&id=81&Itemid=119 (accessed February 3, 2013).

[114] U.S. Department of Commerce, "Memorandum of Understanding between the National Oceanic and Atmospheric Administration of the Department of Commerce of the United States of America and the Russian Academy of Sciences of the Russian Federation on the Cooperation in the Area of the World Oceans and Polar Regions," 2003.

[115] National Oceanic and Atmospheric Administration, Ocean Explorer, "Russian-U.S. Arctic Census 2012," http://oceanexplorer.noaa.gov/explorations/12arctic/welcome.html (accessed March 19, 2013).

created by the Arctic Council. Most recently, the US presented its country report in January 2012, which is one of many reports that SAON uses to develop an overarching report to the Arctic Council.[116] In order to support SOAN, the US has its own Arctic Observing Network (AON). AON is a network of observers such as local citizens, scientists on ships, buoys, and aircraft that report Arctic observations in environmental change.[117] The NSF funds numerous programs, mostly graduate students who conduct studies and participate in AON.[118]

NSPD 66 – HSPD 25 directs the appropriate Departments and Agencies to "[p]romote regular meetings of Arctic science ministers or research council heads to share information concerning scientific research opportunities and to improve coordination of international Arctic research programs." Members of the US Arctic scientific research community, some working with assistance of USG grants or in conjunction with US Agencies, participate in numerous committees, meetings, working groups, both US and international. One example is the Annual Meeting of the International Arctic Buoy Programme (IABP) at the University of Washington.[119] Other examples include Arctic Observing Summit (AOS) and the SAON Board meeting held recently in Potsdam, Germany, where the US presented the latest efforts of the Interagency Arctic Research Policy Committee (IARPC), including its five year plan as discussed below.[120]

---

[116] Sustaining Arctic Observing Networks, National Reports, "US National Report," http://www.arcticobserving.org/national-reports (accessed February 3, 2013).

[117] Study of Environmental Arctic Change, "2012 U.S. Arctic Observing Coordination Workshop," http://www.arcus.org/search/aon (accesses Feb 15, 2013).

[118] National Science Foundation, Arctic Observing Network (AON), "Program Guidelines,' http://www.nsf.gov/funding/pgm_summ.jsp?pims_id=503222&org=NSF (accessed February 12, 2013).

[119] International Arctic Buoy Programme, Polar Science Center, University of Washington, "Overview," http://iabp.apl.washington.edu/index.html (accessed February 11, 2013).

[120] Sustaining Arctic Observing Networks (SAON), "Board Meetings," http://www.arcticobserving.org/news (accessed February 3, 2013).

NSPD 66 – HSPD 25 also requires the agencies to "[w]ork with the Interagency Arctic Research Policy Committee (IARPC) to promote research that is linked strategically to U.S. policies articulated in this directive, with input from the Arctic Research Commission". This requirement is redundant as the agencies tasked to work with the IARPC actually make up the IARPC as evident in their 2012 principles list. Representatives from DOD, DHS, DOS, Department of Commerce (DOC), Department of Energy (DOE), Department of the Interior (DOI), Department of Transportation (DOT), Environmental Protection Agency (EPA), U.S. Department of Health and Human Services (HHS), National Aeronautics and Space Administration (NASA), NSF and others form the IARPC.[121] Additionally, these agencies work closely together to ensure money devoted to Arctic research follows the seven goals listed in IARPC's five-year plan, which are: sea ice and marine ecosystem studies; terrestrial ice and ecosystem studies; atmospheric studies of surface heat, energy, and mass balances; observing systems; regional climate models; adaptation tools for sustaining communities; and human health studies.[122]

Perhaps the strongest and most robust US involvement in Arctic research comes from independent researchers such as graduate students and universities using USG grants. The University of Washington has a Polar Science Center. The University of Colorado, Boulder has an Institute of Arctic and Alpine Research. The International Arctic Research Center is at the University of Alaska Fairbanks.[123] Additionally, the NSF supports numerous Arctic research projects such as the one at the University of Maryland Center for Environmental Science in which scientists deployed aboard the

---

[121] National Science Foundation, "Interagency Arctic Research Policy Committee Principals 2012," IARPC Principals list, http://www.nsf.gov/od/opp/arctic/iarpc/iarpc_principals2012.jsp (accessed February 3, 2013).

[122] Interagency Arctic Research Policy Committee, *Arctic Research Plan: FY2013-2017,* September, 2012; and Ibid., 3.

[123] National Oceanic and Atmospheric Administration, "Research institutions and organizations focused on the Arctic," http://www.arctic.noaa.gov/orgs.html (accessed February 3, 1013).

USCGC HEALY and USCGC POLAR SEA.[124] With all these different universities, both in the US and abroad, it became apparent that collaboration of research was necessary to document fully the tremendous changes occurring in the Arctic. During the mid-1990s, a professor from the University of Washington's Polar Science Center began reaching out to other institutions and individual scientists to form a group that later became the Study of Environmental Arctic Change (SEARCH) program. With the NSF's backing, SEARCH became an internationally recognized Arctic research program.[125] Together these programs have far surpassed the tasking in NSPD 66 – HSPD 25 to "[s]trengthen partnerships with academic and research institutions and build upon the relationships these institutions have with their counterparts in other nations." [126]

Overall, the scientific community is doing an outstanding job at implementing US policy. In this arena the US government did not lead the way, yet, the Government eagerly embraced opportunities presented from universities and independent research scientists and provided grant money.[127] As the Arctic rapidly changes, so too does the nation's academic community. Keeping the US active in the international arena regarding scientific research and cooperation requires scientists to deploy to the Arctic, often aboard an icebreaker.

Maritime Transportation in the Arctic Region

With the possibility of three new Arctic routes becoming a reality in the near future, the USG

---

[124] University of Maryland, Center for Environmental Science Chesapeake Biological Laboratory. "Bering Ecoystem Study (BEST) and Bering Sea Integrated Ecosystem Research Program (BSIERP)". http://arctic.cbl.umces.edu/ (accessed February 3, 2013).

[125] Study of Environmental Arctic Change (SEARCH), "Development of SEARCH," "1990-1999: Early Activities," http://www.arcus.org/search/sciencecoordination/development (accessed February 3, 2013).

[126] President, *National Security Presidential Directive/NSPD 66 – Homeland Security Presidential Directive /HSPD 25* (January 9, 2009). For the text of NSPD 66/HSPD 25, 5.

[127] National Science Foundation, "Arctic Research and Education," http://www.nsf.gov/funding/ pgm_summ.jsp?pims_id=13448 (accessed February 3, 2013).

Policy sought new measures to deal effectively with issues that may arise from increased shipping in the Arctic. It specifically called for increased cooperation with other nations. In November, 2004, the Arctic Council tasked the Council's Protection of the Arctic Marine Environment (PAME) sub-group to undertake an assessment on Arctic marine shipping. This assessment, published in 2009 as *The Arctic Marine Shipping Assessment (AMSA),* focused on impacts made to the Arctic by ships. This included impacts on humans, the environment, and the infrastructure needed to facilitate Arctic marine transportation. The assessment required input from various organization such as shipping companies, insurance companies, ship architects, shipping associations, and classification organizations.[128]

In NSPD 66 – HSPD 25, the President tasked the Secretaries of State, Defense, Transportation, Commerce, and Homeland Security with implementing USG policy regarding maritime transportation in the Arctic. AMSA was the first step to implementing USG policy as it helped, "[d]evelop additional measures, in cooperation with other nations, to address issues that are likely to arise from expected increases in shipping into, out of, and through the Arctic region".[129] AMSA provided a baseline assessment of the status of maritime transportation in the Arctic.

The implementation plan also calls for agencies to "[c]ommensurate with the level of human activity in the region, establish a risk-based capability to address hazards in the Arctic environment".[130] Although the USCG quotes this requirement in its Arctic strategic approach and states that it will be a leader in implementing national policy it has not yet developed a risk-based capability.[131] The USN, in its Arctic Roadmap, like the USCG, quotes the requirement, but never again uses the phrase risk-based when

---

[128] Arctic Council, *Arctic Marine Shipping Assessment 2009 Report*, April 2009.

[129] President, *National Security Presidential Directive/NSPD 66 – Homeland Security Presidential Directive /HSPD 25* (January 9, 2009). For the text of NSPD 66/HSPD 25, 6.

[130] Ibid., 6.

[131] U.S. Coast Guard. Commandant Instruction 16003.1, *U.S. Coast Guard Arctic Strategic Approach* (Washington, DC, April 26, 2011).

dealing with Arctic capabilities.[132] DoD's 2011 report to Congress, used risk-based analysis to say that DoD felt that a low risk to National Security currently existed in the Arctic, and that a risk-based strategy was necessary in procurement of ships.[133]

The USG policy calls for a risk-based assessment to "advance work on pollution prevention and response standards."[134] Currently, the International Convention for the Prevention of Pollution from Ships (MARPOL) is the primary governing body regarding pollution from maritime transportation.[135] In the US, the National Response System is the command and control system for oil spills. For oil spills occurring in US waters of the Arctic, the USCG is the lead agency.[136] As part of Arctic Shield 2012 the USCG conducted several oil spill response drills including exercising the oil skimming capabilities of its Juniper Class cutters, which although primarily designed for buoy tending, was also constructed with the ability to deploy skimming equipment as well as the ability to store the collected oil.

---

[132] U.S. Department of the Navy, *U.S. Navy Arctic Roadmap* (Washington, DC, October 2009). 1-33.

[133] U.S. Department of Defense, Report to Congress on Arctic Operation and the Northwest Passage. (Washington, DC, May 2011), 19.

[134] President, *National Security Presidential Directive/NSPD 66 – Homeland Security Presidential Directive /HSPD 25* (January 9, 2009). For the text of NSPD 66/HSPD 25, 6.

[135] International Maritime Organization. "International Convention for the Prevention of Pollution from Ships (MARPOL)," http://www.imo.org/About/Conventions/ListOfConventions/ Pages/International-Convention-for-the-Prevention-of-Pollution-from-Ships-(MARPOL).aspx (accessed Feb 4, 2013).

[136] U.S. Coast Guard, National Response Center "National Response System," http://www.nrc.uscg.mil/nrsinfo.html (accessed Feb 4, 2013).

Figure 7. USCGC SYACMORE deploys the Spilled Oil Recovery System (SORS) during an exercise near Barrow, Alaska, July 31, 2012 using a barge with tanks to store recovered oil.

Source: U.S. Coast Guard photo by Petty Officer 2nd Class Kelly Parker.

In May 2011, the Arctic Council formed a task force on Arctic Marine Oil Pollution Preparedness and Response. The task force is co-chaired by US Ambassador David A. Balton who is the Deputy Assistant Secretary for Oceans and Fisheries in the Department of State's Bureau of Oceans, Environment and Science. [137] At a meeting on October 11, 2012 in Reykjavik, Iceland the task force developed a draft international agreement which, once signed, is a legally binding agreement, but also contains non-legally binding appendixes.[138] The agreement provides requirements regarding maintaining national systems for oil pollution preparedness and response, notification of other states, monitoring, requests for assistance and coordination and cooperation in response operations' movement and removal of resources across

[137] Julia L. Gourley, "Pressing Issues in the Arctic" (Department of State PowerPoint Presentation given September, 7, 2011, Anchorage, AK) http://onlinepubs.trb.org/onlinepubs/mb/2011Fall/ppt/1gourley.pdf (accessed Feb 4, 2013); and Environmental Law Institute, "David A. Bolton Biography," http://www.eli.org/pdf/ocean/seminars/balton_bio.pdf (accessed Feb 4, 2013).

[138] "Agreement on Arctic Oil Spill Close to Completion," *Iceland Review* Online November 12, 2012.http://www.icelandreview.com/icelandreview/daily_news/Agreement_on_Arctic_Oil_Spill_Close_to_Completion_0_394301.news.aspx (accessed 6 Mar 2013).

borders, joint review of oil pollution incident response operations, cooperation and exchange of

information, and joint exercises and training. [139] Fortunately, the US already has a national system in

place, however, the US Arctic infrastructure is nearly non-existent and multiple surface ships will be

required to travel great distances just to get on scene and resupplying these vessels will be extremely

difficult.

Incidentally, the Presidents next tasking was to "determine basing and logistics support

requirements." [140] USCG Captain Peter Troedsson, currently a Military Fellow at the Council on Foreign

Relations and most recently the Chief of Staff of the Eighth Coast Guard New District during the

Deepwater Horizon Oil Spill in the Gulf of Mexico, easily identified the gaps in his article in Fletcher

Forum:

> Find Kaktovik, Alaska on a map and take note of the road system, or the next nearest town.
> Transportation of personnel and equipment, berthing, food, water, shelter, decontamination, and
> communications capabilities in these remote areas would be a monumental challenge for a large
> scale response operation. Port facilities in the area can accommodate only shallow draft vessels,
> and airfields have only short, gravel runways. A lack of road systems and a complete dearth of
> hotels for lodging and staging capability complete the picture. A significant investment in
> infrastructure is needed. [141]

Although DoD's Arctic Report to Congress does assess existing Arctic infrastructure for national security

purposes it does not address required infrastructure for an oil spill response. [142] Whereas the USCG can

---

[139] Draft Arctic Council Agreement as released by Greenpeace, "Agreement on Cooperation on Marine Oil Pollution Preparedness and Response in the Arctic," 11 October 2012 http://www.greenpeace.org/international/Global/international/briefings/climate/ArcticCouncilDraftAgreementCooperationOPPR.pdf (accessed 6 Mar 2013).

[140] President, *National Security Presidential Directive/NSPD 66 – Homeland Security Presidential Directive /HSPD 25* (January 9, 2009).

[141] Peter Troedsson, Capt USCG, "Leading Preparedness for an International Oil Spill Response in the Arctic," *The Fletcher Forum of World Affairs* (February 2013) http://www.fletcherforum.org/2013/02/20/troedsson/ (accessed 6 Mar 2013).

[142] U.S. Department of Defense, *Report to Congress on Arctic Operation and the Northwest Passage*. (Washington, DC, May 2011), 3.

easily point out the enormous lack of infrastructure in the Arctic, it has not yet identified the minimum

infrastructure needed to support a large oil spill recovery operation in the Arctic.

Figure 8. The USCG rented facility in Barrow, Alaska that served as its Forward Operating Location during Arctic Shield 2012, along with two USCG MH-60 Jayhawk helicopters deployed from Air Station Kodiak, July 20, 2012.

Source: U.S. Coast Guard photo by Petty Officer 2nd Class Elizabeth H. Bordelon.

However, data "including necessary airlift and icebreaking capabilities" has partially been

identified. The US Government Accountability Office (GAO) reported to a Congressional Committee

regarding Arctic Capabilities stating that DoD only partially addressed the need for icebreaking

capabilities in its Arctic Report to Congress, but that the USCG's High Latitude Study gave specific

requirements for icebreakers needed to carry out the Coast Guard's missions, one of which is oil spill

prevention and response. [143] Regarding airlift, DoD's Arctic Report specifically states that the region,

---

[143] U.S. Government Accountability Office, Report to Congressional Committees, *Arctic Capabilities: DOD Addressed Many Specified Reporting Elements in Its 2011 "Arctic Report" but Should Take Steps to Meet Near-and-Long-term Needs*, GAO -12-180, 28.

including airfields are adequate for US defense needs, but again there have been no efforts to identify the airlift needs for a major oil spill response.[144]

Another HSPD 66/HSPD 25 tasking in this section was "improve plans and cooperative agreements for search and rescue." [145] On May 12, 2011, Secretary of State Hillary Clinton, representing the US as part of the Arctic Council, signed an agreement obligating the US to provide Maritime Search and Rescue (SAR) in a designated region of the Arctic.[146] In preparations for signing of the treaty, the USCG's capabilities to operate in the Arctic became a major focus, but in a hearing on the USCG's proposed FY2012 budget before a Senate Subcommittee, Admiral Robert Papp, Commandant of the USCG stated, "we've got zero capability to respond in the Arctic right now." [147]

Figure 8. USCGC BERTHOLF provides limited summer Arctic SAR capability with an embarked MH-

---

[144] U.S. Department of Defense, *Report to Congress on Arctic Operation and the Northwest Passage.* (Washington, DC, May 2011), 3.

[145] President, *National Security Presidential Directive/NSPD 66 – Homeland Security Presidential Directive /HSPD 25* (January 9, 2009).

[146] Congressional Research Service, *Changes in the Arctic: Background and Issues for Congress* by Ronald O'Rourke. Washington, DC, 2013, 41.

[147] U.S. Congress. Senate Committee on Commerce, Science, and Transportation. *Testimony of Admiral Robert Papp Commandant, U.S. Coast Guard on "Arctic Operations": Hearing before the Subcommittee on Oceans, Atmosphere, Fisheries, and Coast Guard.* 112 Cong. 1st sess., June 23, 2011.

65 Dolphin helicopter, Sept. 14, 2012 as part of Arctic Shield 2012.

Source: U.S. Coast Guard photo by Petty Officer 1st Class Timothy Tamargo.

The USG policy next addressed Arctic waterways and safe navigation. It specifically tasked agencies to "[d]evelop Arctic waterways management regimes in accordance with accepted international standards, including vessel traffic-monitoring and routing."[148] The USCG is the federal agency responsible for vessel traffic services (VTS), yet the USCG has no VTS operational in the Arctic or the Bering Strait, which would be exceptionally expense to build and operate. However, the USCG can track vessels via the Automated Identification System (AIS), which is required on most commercial vessels operating in the Arctic.[149] NSPD 66/HSPD25 also calls for "safe navigation standards." As previously, discussed COLREGS are the international navigation standards and apply to Arctic shipping.[150] In June 2011, NOAA issued the Arctic Nautical Charting Plan to focus on improving existing and developing new nautical charts to cover the expanding maritime routes in the Arctic.[151] There are 568,000 square nautical miles (SNM) in the U.S. Arctic Exclusive Economic Zone.[152] Both the USCG and NOAA are taking step to provide "accurate and timely environmental and navigational information." For NOAA this includes conducting surveys on what it calls navigationally significant areas, which could take

---

[148] President, *National Security Presidential Directive/NSPD 66 – Homeland Security Presidential Directive /HSPD 25* (January 9, 2009), 6.

[149] Title 33, codified at *U.S. Code* 33 (2003), § 164046.

[150] International Maritime Organization, *Convention on the International Regulations for Preventing Collisions at Sea, 1972*, codified at 33 USCS § 1602.

[151] National Oceanic and Atmospheric Administration, "Arctic Navigation; How is NOAA improving the nautical charts ships need?" http://oceanservice.noaa.gov/economy/arctic/#4 (accessed Mar 8, 2013).

[152] National Oceanic and Atmospheric Administration, "Arctic Navigation; What about hydrographic surveying for navigation?" http://oceanservice.noaa.gov/economy/arctic/#5 (accessed Mar 8, 2013).

approximately twenty-five years given current resources.[153] Through Arctic Shield 2012, the USCG continued to assess capability gaps and provided the summer time capability to fulfill its missions in the Arctic. Regarding the requirement to "[e]valuate the feasibility of using access through the Arctic for strategic sealift and humanitarian aid and disaster relief,"[154] it appears that no department or agency has fully conducted such an assessment.

Energy

Energy development in the Arctic has the potential to play an enormous role in future world economics as more nations and companies begin to explore the vast amount of oil, gas, and minerals estimated to be in the Arctic. The US policy focuses on protecting its interests, while fostering partnerships, both domestic and international, with other nations, universities, regulating authorities and organizations in conducting research and exploration for energy in the Arctic. The policy tasked the Secretaries of State, the Interior, Commerce, and Energy to increase study efforts by working with Arctic nations and partners on numerous issues.[155] In September 1998, the Arctic Council chartered the Sustainable Development Working Group (SDWG) to, among other things, "propose and adopt steps to be taken by the Arctic States to advance sustainable development in the Arctic, including opportunities to protect and enhance the environment and the economies."[156] The US representative to the SDWG during the time was Karen Perdue, Commissioner of the Alaska Department of Health and Social Services who

---

[153] Ibid.

[154] President, *National Security Presidential Directive/NSPD 66 – Homeland Security Presidential Directive /HSPD 25* (January 9, 2009), 6.

[155] President, *National Security Presidential Directive/NSPD 66 – Homeland Security Presidential Directive /HSPD 25* (January 9, 2009).

[156] Arctic Council, *Sustainable Development Program (SD Program)*, http://www.sdwg.org/content.php?doc=12 (accessed 16 Mar 2013).

has no experience with energy issues.[157] Prior to the release of HSPD 66/NSPD 25 the State Department sponsored the Arctic Energy Summit organized by the Institute of the North.[158] Panelists and presenters from the US included representatives for the Departments of State, Interior, Energy, and Homeland Security, as well as representatives from the State of Alaska, numerous non-profit agencies, universities, industry, local governments and tribal organizations. The final report of the Arctic Energy Summit shows the depth of cooperation in energy issues regarding the Arctic.[159] Then in 2010, the SDWG released its *Report on Arctic Energy* that serves as a background paper presenting an overall look at energy in the Arctic.[160] The paper points to the Arctic Energy Summit as the US's big contribution to the project. Intended for a multinational audience, the paper presents very little recommendations other then continued summits, conferences, and overall international cooperation. Although the most recent reports were not released until 2010, the actual activities in both the summit and the working group took place before the US policy was issued.

Another controversial area regarding US energy in the Arctic revolves around government oil and gas leasing. In 2009, the USG did not extend the congressional moratorium on oil and gas leasing activities as it had since 1982, effectively opening up the US continental shelf for energy exploration.[161] The Bureau of Ocean Energy Management (BOEM), which resides within the US Department of the Interior, is the government agency responsible for oil and gas energy programs. On August 27, 2012, the

---

[157] Alaska state Hospital and Nursing Home Association, http://www.ashnha.com/about/contact-us/ (accessed March 16, 2013) and Arctic Council, *Sustainable Development Working Group Report on Arctic Energy*, (2009), 34.

[158] The Institute of the North is a non –profit organization founded by Alaskan Governor Walter J. Hickel to conduct research in the north to include the Arctic.

[159] Arctic Council. *The Arctic Energy Summit Final Report and Technical Proceedings*, prepared by James R. Hemsath. Anchorage: Institute of the North, 2010.

[160] Ibid., 34.

[161] Congressional Research Service, *Changes in the Arctic: Background and Issues for Congress* by Ronald O'Rourke. Washington, DC, 2011, 20.

BOEM released the Five Year Outer Continental Shelf (OCS) Oil and Gas Leasing Program. This leasing program establishes the areas that the USG thinks is best suited to meet national energy needs and then leases these areas to commercial enterprises for exploration and drilling. The five-year plan identifies 125.19 million acres for lease in the US Arctic. [162] And in December 16, 2011 BOEM issued conditional approval for Shell Oil to begin drilling in the Chukchi Sea during the summer of 2012.[163] BOEM also has existing leases for the Chukchi Sea and Cook Inlet in 2016 and the Beaufort Sea in 2017.[164]

Figure 9. Shell Oil's 514-foot drill ship Noble Discoverer 68 miles west of Nome, Alaska. *Source:* U.S. Coast Guard photo by Air Station Kodiak.

For the Department of Interior, the US Geological Survey (USGS) estimated vast amounts of natural energy in the Arctic in 2008, and has since focused mainly on defining the outer limits of the U.S. continental shelf and conducted research in non-energy- related exploration such as the Arctic ecosystem

---

[162] Bureau of Ocean Energy Management, "Five Year Outer Continental Shelf (OCS) Oil and Gas Leasing Program," http://www.boem.gov/5-year/2012-2017/ (Accessed March 17, 2013).

[163] Bureau of Ocean Energy Management, "Shell 2012 Exploration Plan – Chukchi Sea," http://www.boem.gov/ShellChukchi2012/ (Accessed March 17, 2013).

[164] Bureau of Ocean Energy Management, "2012- 2017 Lease Sale Schedule", http://www.boem.gov/Oil-and-Gas-Energy-Program/Leasing/Five-Year-Program/Lease-Sale-Schedule/2012---2017-Lease-Sale-Schedule.aspx (accessed March 17, 2013).

and marine mammals. USGS launched a new research effort on August 25, 2012, to study acidification in the Arctic as well as continued efforts to define the US continental shelf aboard USCGC HEALY.[165]

Perhaps the busiest agency regarding energy in the Arctic was the National Oceanic and Atmospheric Administration (NOAA) within the Department of Commerce. Starting in 2010, the NOAA ship FAIRWEATHER has been underway in the Arctic conducting hydrographic surveys in order to chart more effectively the Arctic sea floor.[166] And in December 2011, NOAA released the "Draft Environment Impact Statement (DEIS) for the Effects of Oil and Gas Activities in the Arctic Ocean," which analyzes the effects that offshore oil and gas activities could have on various Arctic species, resources, and indigenous people. [167] Not to be left out, in May 2012, the Energy Department announced that through cooperation with Japanese and US Oil companies it had successfully tested methods to extract natural gas from methane hydrates on Alaska's North Slope.[168]

Environmental Protection and Conservation of Natural Resources

The USG policy next asserts that the Arctic is a sensitive environment that is not yet fully understood and states that more scientific studies need to be conducted. It specifically stresses the need for studies regarding the ecosystem in order to develop an effective long-term plan for the Arctic in regard to managing Arctic species. Vis-à-vis implementation, the policy requires the Secretaries of State,

---

[165] Arctic Update; The US Arctic Research Commission Daily Email Newsletter, August 28, 2012, http://www.arctic.gov/arctic_update_archive/2012aug28.html (accessed March 17, 2013).

[166] U.S. National Oceanic and Atmospheric Administration. "NOAA Ship Fairweather sets sail to map areas of the Arctic," (Washington, DC, July 7, 2011) http://www.noaanews.noaa.gov/stories2011/20110707_fairweather.html (accessed March 17, 2013).

[167] Department of Commerce, National Oceanic and Atmospheric Administration, "Notice of Availability of a Draft Environmental Impact Statement for the Effects of Oil and Gas Activities in the Arctic Ocean," *Federal Register* 76, no. 251 (December 30, 2011), 82275.

[168] U.S. Department of Energy, "On the Frontiers of a New Energy Source,"(Washington, DC, May 2, 2012) http://energy.gov/articles/frontiers-new-energy-source (accessed March 17, 2013).

the Interior, Commerce, and Homeland Security as well as the Administrator of the Environmental

Protection Agency to work with other nations to "respond effectively to increased pollutants and other

environmental challenges".[169]

The Magnuson-Stevens Fishery Conservation and Management Act (MSFCMA) created eight

Regional Fishery Management Councils (FMC) to manage fisheries in the US Exclusive Economic Zone

(EEZ). NOAA oversees these councils and is responsible for approving the management plans

recommended by each FMC.[170] The North Pacific FMC recommended a plan in February 2009 that

prohibits commercial fishing in the Arctic, stating it needs more information to develop a sustainable

fisheries management plan for the Arctic. In August 2009, the Secretary of Commerce announced the

"Fisheries Management Plan for the Fish Resources of the Arctic Management Area" that closed a large

section of the Arctic to commercial fishing and changed the King and Tanner Crab fishing regulations for

the Bering Sea and Aleutian Islands. The plan affords for a study of the "Arctic's fish resources and their

habitat (including essential fish habitat definitions), current fishing activities, the economic and

socioeconomic characteristics of current fisheries and communities, and ecosystem characteristics" as was

required by the USG policy.[171]

The policy also mandates the departments to identify ways to "ensure adequate enforcement

presence" for safeguarding Arctic marine life. The Magnuson-Stevens Fisheries Conservation and

Management Act (MSFCMA) tasks the USCG with enforcement of fisheries laws at sea.[172] As such, each

---

[169] President, *National Security Presidential Directive/NSPD 66 – Homeland Security Presidential Directive /HSPD 25* (January 9, 2009).

[170] Department of Commerce. Public Law 94-265 Magnuson-Stevens Fishery Conservation and Management Act. U.S. Department of Commerce, National Oceanic and Atmospheric Administration, National Marine Fisheries Service January 12, 2007, 63.

[171] North Pacific Fishery Management Council ,"Fishery Management Plan for Fish Resources of the Arctic Management Area", Anchorage, Alaska, 2009. ES-4. https://alaskafisheries.noaa.gov/npfmc/ PDFdocuments/fmp/Arctic/ArcticFMP.pdf (accessed March 18, 2013).

[172] Department of Commerce. Public Law 94-265 Magnuson-Stevens Fishery Conservation and

Regional Fishery Management Council has a non-voting USCG member.[173] The USCG also works closely with the State Department to develop and enforce international fishing agreements. The USCG released its ten year Fisheries Enforcement Strategic Plan, *Ocean Guardian* in 2004, which lays out the Coast Guard's enforcement role "in support of the national goals for fisheries resource management and conservation." [174] *Ocean Guardian* explains the linkage between NOAA, the State Department, and the USCG strategic plans regarding marine fisheries law enforcement. The USCG focuses on four key concepts: sound regulations, effective presence, application of technology, and productive partnerships, with effective presence defined as "the allocation of fisheries enforcement resources at levels that ensure adequate compliance with management measures implemented to recover and maintain healthy fish stocks."[175] However, *Ocean Guardian* never mentions the USCG's limited ability to provide effective presence in the Arctic.

Figure 10. USCGC POLAR SEA conducts a fisheries patrol off Kodiak, Alaska, April 9, 2008.

Management Act. U.S. Department of Commerce, National Oceanic and Atmospheric Administration, National Marine Fisheries Service, January 12, 2007, 47.

[173] Ibid., 64.

[174] U.S. Coast Guard, *Ocean Guardian,* September 20, 2004. Washington, DC, 1.

[175] Ibid., 5.

Figure 11. USCGC SHERMAN conducts a fisheries boarding as a crewmember on a fishing vessel prepares to set a crab pot in the Bering Sea.

Source: Coast Guard photo by Petty Officer 3rd Class Erik Swanson.

The USG Policy also states, "the United States shall continue to collaborate with other governments to ensure effective conservation and management." Specifically regarding fisheries, the USG Policy encourages "international agreements or organizations to govern future Arctic fisheries." [176] Actually, in 2008 President George W. Bush signed a resolution:

> Resolved by the Senate and House of Representatives of the United States of America in Congress assembled, That (1) the United States should initiate international discussions and take necessary steps with other Arctic nations to negotiate an agreement or agreements for managing migratory, transboundary, and straddling fish stocks in the Arctic Ocean and establishing a new international fisheries management organization or organizations for the region.[177]

Yet as of 2013, there is still no multinational agreement or organization specifically for managing commercial fishing in the Arctic Ocean, and the Arctic Council has not yet created a working group for

---

[176] President, *National Security Presidential Directive/NSPD 66 – Homeland Security Presidential Directive /HSPD 25* (January 9, 2009).

[177] U.S. Congress. Senate. *A joint resolution directing the United States to initiate international discussions and take necessary steps with other nations to negotiate an agreement for managing migratory and transboundary fish stocks in the Arctic Ocean.* S. 17. 110th Cong., 2d sess., (May 23, 2008), 2.

commercial Arctic fisheries. [178]

The USG Policy also calls for intensified study of "adverse effects of pollutants on human health" in the Arctic.[179] The Arctic Council's AMAP working group released numerous reports in 2009 to include a report on human health in the Arctic, which discusses, "how global climate change and climate variability, global and regional control initiatives, local industrial activities and social and cultural activities may be influencing environmental contamination and/or human exposure and vulnerability in the Arctic."[180]

RESOURCES, ASSETS, AND FUNDING

The USG Policy concludes with a paragraph discussing the resources and assets required to implement national strategy in the Arctic tasking Departments and Agencies to identify future funding needs and if necessary proceed with legislative proposals.[181] A quick look at fiscal year 2013 proposed budgets and historical congressional actions prove that the Arctic is not a priority for the US. The State Departments 2013 budget includes $32,800,000 for international fisheries. Of that, only $58,000 is earmarked for the Arctic, specifically for the Arctic Council. The State Department also requested $2,526 million for development assistance to support the Presidential Policy Directive on Global Development (PPD-6). The $2,526 million does provide development assistance for indigenous communities due to

---

[178] U.S. Department of State, Treaties in Force: A List of Treaties and Other International Agreements of the United States in Force on January 1, 2012 available www.state.gov/documents/organization/202293.pdf (accessed March 17, 2013).

[179] President, *National Security Presidential Directive/NSPD 66 – Homeland Security Presidential Directive /HSPD 25* (January 9, 2009).

[180] AMAP Assessment 2009: *Human Health in the Arctic.* Arctic Monitoring and Assessment Programme (AMAP), Oslo, 2009, 9.

[181] President, *National Security Presidential Directive/NSPD 66 – Homeland Security Presidential Directive /HSPD 25* (January 9, 2009).

global climate change and includes agricultural productivity, fisheries, public health and energies. [182] However, it is unclear the dollar amount allocated to the Arctic indigenous people.

In January 2012, DoD published two key documents the explains their financial strategy; *Sustaining US Global Leadership: Priorities for 21st Century Defense* and *Defense Budget Priorities and Choices (P21)*. [183] Neither document specifically mentions the Arctic or global climate change. However, the both documents do discuss the necessity that the "United States must maintain its ability to project power".[184] Yet, DoD's 2013 budget has no request for new surface vessels capable of operating year round in the Arctic.[185] The Navy says it will address Arctic funding in 2014, and DOD's 2013 budget does include additional monies for Ballistic Missile Defense (BMD) both afloat and ashore including the Alaska site.[186] In fact, the House-approved version substantially increases the BMD funding.[187]

Addressing DHS's role in national security, Secretary Janet Napolitano's 2013 budget press release stated "[i]n addition to supporting Coast Guard's current operations in the Polar Regions, the budget initiates acquisition of a new polar icebreaker to address Coast Guard emerging missions in the Arctic." [188] Additionally, the 2013 Coast Guard budget enacted by the Senate and House of

[182] U.S. Department of State http://www.state.gov/documents/organization/183755.pdf Executive Budget Summary *"Function 150 & Other International Programs Fiscal Year 2013"*, 58 and 81.

[183] Department of Defense. *"Overview"* Office Of The Under Secretary Of Defense (Comptroller) / Chief Financial Officer, February 2012, preface.

[184] Department of Defense. *Sustaining US Global Leadership: Priorities for 21st Century Defense.* Washington, DC. January 5, 2012, 4.

[185] Pat Towell and Daniel H. Else. CRS Report. "Defense: FY2013 Authorization and Appropriations." September 5, 2012. Washington, DC, 1-50.

[186] U.S. Department of the Navy, *U.S. Navy Arctic Roadmap* (Washington, DC, October 2009), 4.

[187] Ibid., 23-24.

[188] Department of Homeland Security *"Secretary Napolitano Announces Fiscal Year 2013 Budget Request"* (Washington, DC, February 13, 2012). http://www.dhs.gov/news/2012/02/13 /secretary-napolitano-announces-fiscal-year-2013-budget-request (accessed March 19, 2013).

Representative on January 3, 2012, requires the USCG to provide an analysis on options and cost for the reactivation of USCGC POLAR SEA for an additional ten years. The USCG budget also provides for the establishment of an Integrated Cross-Border Maritime Law Enforcement Operations Program with Canada.[189] The appropriations bill also requires that USCG to submit an assessment of additional USCG needs in the Arctic to include shore infrastructure in areas that are ice-free year round. The assessment is required to include the capabilities of all USCG assets other than icebreakers. The appropriations bill also tasks the USCG, US Navy and US Army Corps of Engineers with conducting a study on the establishment of a deep water seaport in the Arctic to "protect and advance strategic United States interests within the Arctic region."[190]

This active interest buy Congress seems hopeful; however, several agencies have been submitting studies to Congress for years stating the status of the US's icebreakers without an attendant increase in funding. Meanwhile the Customs and Border Protection (CBP) portion of DHS's budget included only $10 million for enhancement on the Northern border technologies.[191] NOAA's 2013 budget provide $8 million to research the changing climate processes including the Arctic, and also contains grant monies to fund marine research on fisheries or marine ecosystems in the Arctic via the Environmental Improvement and Restoration Fund. [192] The Department of Interior's Bureau of Ocean Energy Management (BOEM) continued to receive funding for a five year study of Hanna Shoal ecosystem in the Chukchi Sea.[193] DOI's

---

[189] U.S. Congress 112th 2nd session. HR 2838. Appropriations for the Coast Guard for fiscal year 2013 through 2014, 42.

[190] Ibid., 44.

[191] Department of Homeland Security *FY 13 Budget-in-Brief,* Washington, DC. Undated. p. 87. http://www.dhs.gov/xlibrary/assets/mgmt/dhs-budget-in-brief-fy2013.pdf (accessed March 19, 2013).

[192] U.S. Department of Commerce *Budget in Brief Fiscal Year 2013,* John E. Bryson, Secretary Washington, DC, 6; and Ibid., 71.

[193] U.S. Department of Interior, *Fiscal Year 2013 The Interior Budget in Brief,* Washington, DC February 2012, DO-9.

2013 budget also contains $14.9 million for oil spill research including research in Arctic environments.[194] An additional $500,000 was identified for research efforts through the DOI Climate Science Centers to enhance work with Tribes to understand the impacts of climate change on tribal lands.[195]

Given the above activity, it appears as if the United States has taken a keen interest in the Arctic for FY 2013. However, at $613.9 billion, DoD's 2013 proposed budget is 4.3 times larger than DoS, DHS, DoI, DoC, DoE, and HHS's combined 2013 budgets. Truth be told, the US is not putting forth the funding necessary to get on top of the emerging changes in the Arctic. And although, funding for one ice breaker is a step in the right direction it does little to alleviate the pressures that currently fall on one icebreaker. And, given sequestration and a continuing resolution funding for the above programs may not come to fruition.

CONCLUSION

The Arctic is melting and few in Washington seem to be paying attention. There are varied US national interests in the Arctic including national and homeland security, international governance, continental shelf and boundary issues, promotion of international scientific cooperation, Arctic maritime transportation, economic issues, and environmental protection and conservation of natural resources. And there are as many varied departments, agencies, universities, and government and non-government organizations developing policies and conducting assessments and studies which seem to be all but ignored by Congress.

The President made it clear in NSPD 66/HSPD 25 that national defense was the nation's number one priority and for the Arctic, that largely fell to the Navy and Coast Guard as the Arctic is primarily a maritime environment. All departments and agencies agree that freedom of navigation and the ability to

---

[194] Ibid., BH-27.

[195] Ibid., DO-15.

project sea power are critical to the US national defense. DoD and the US Navy both admit that they do not have the capability to fulfill this requirement, yet neither are taking any action to build surface ship capability in the Arctic. DoD and the Navy state that the USCG is responsible for maintaining the nation's icebreaker fleet and seemingly relieve themselves of their duty to ensure freedom of navigation and assured access to all international waters. Unfortunately, Washington continues to sit idly on the sideline requesting additional studies, research and assessments, even though almost all departments have clearly made a case for additional icebreakers and the need for the US to project sea power via surface ships. Even with funding for an icebreaker in the Coast Guard's 2013 budget, Congress requested additional studies be done before providing any funding to restore USCGC POLAR SEA. The European Office of the Center for Strategic and International Studies adequately assessed the ongoing situation:

> Unable to make difficult future budget decisions in a constrained budget environment, Washington reverts to a near-constant assessment process of U.S. infrastructure and security needs in the Arctic, suggesting that an endless assessment process is equivalent to taking decision on a future course of action.[196]

In regards to international governance, the US continues to work within many international organizations pertaining to the Arctic and various departments and agencies work hand-in-hand daily with international partners in the Arctic; yet, the results are mostly reports and assessments with recommendations for voluntary compliance. However, Congress needs to understand that the principle of international governance applies to the ratification of UNCLOS. Otherwise, the US is making a statement to the world, that the US is only interested in international agreements when it favors the US and allows the US complete freedom. The same holds true for continental shelf and boundary issues. US scientists continue to conduct research and US vessels continue to complete surveys, yet Congress repeated fails to ratify UNLCOS, despite repeated Presidential, departmental and agency requests to do so, making it impossible for the US to serve as an international leader in the Arctic and impossible to submit a

---

[196] Heather Conley, Terry Toland, and Jamie Kraut. *A New Security Architecture for the Arctic: An American Perspective*. (Washington, DC: CSIS, January 2012), 18.

continental shelf claim. The US failure to ratify UNCLOS also sets precedence for the Russian Federation to claim unilaterally, the Lomonosov Ridge, if their claim under UNCLOS is not accepted, which is in concert with US assertion of continental shelf rights prior to UNCLOS.

Additionally, Washington seems content to 'agree to disagree' with Canada regarding territorial disputes. Ignoring the problem will not make it go away, yet Washington has taken no action to rectify the disputes with Canada or pressure the Russian Federation to approve the agreed-upon treaty. Given Congress' inability to ratify UNCLOS, the US has no clout to assert in getting the Russian Duma to act.

The limited number of research grants that the US gives annually is helping to gain a better understanding of the Arctic; however, the small dollar amount and the lack of US research ships capable of working in the Arctic are themselves an indication that Washington is not concerned about scientific research in the Arctic. With the potential for increased Arctic maritime transportation, the US should be looking at ways to embrace this economies provided by these shorter shipping routes, instead the US is content to let the Canadians take the lead on opening a vessel traffic service, which also affords the Canadians a more robust Arctic maritime domain awareness. Additionally, the US entered into a SAR agreement within the Arctic Council framework that it cannot possibly fulfill, and yet Congress is doing little to increase the USCG's SAR capability in the Arctic. To the credit of the USCG it has attempted to uphold the US agreement during the summer months via its Arctic Shield operations.

Considering economic issues, the USG took steps to protect its Arctic fisheries by closing off certain area to fishing in order to conduct more research before allowing fishing, yet was quick to open its Arctic waters to drilling. The act to open drilling was actually a failure to extend the congressional moratorium on oil and gas leasing activities, leaving a question as to whether it was a deliberate decision or a simple omission. Increased drilling in the Arctic directly relates to the environmental protection and conservation of natural resources. And even though Shell Oil had to pass numerous government requirements regarding oil spill response, the USG's lack of shore infrastructure, oil spill response capabilities, and surface ships capable of operating in the Arctic demonstrates a total lack of regard for the

Arctic environment.

Overall, it is alarming that, as the greatest nation is the world, "[t]he United States today funds a navy as large as the next 17 in the world combined, yet it has just one seaworthy oceangoing icebreaker—a vessel that was built more than a decade ago and that is not optimally configured for Arctic missions."[197] However, the most alarming phenomenon is that the US Government is taking little action to increase its presence in the Arctic or to implement any of the six overarching elements of its national policy, which leaves the US militarily, politically, environmentally, and economically vulnerable in the Arctic.

---

[197] Scott G. Borgerson, "Arctic Meltdown: The Economic and Security Implications of Global Warming," *Foreign Affairs*, vol 87, no. 2. (March/April 2008), 64.

APPENDIX A: PRESIDENT, NATIONAL SECURITY PRESIDENTIAL DIRECTIVE/NSPD 66 – HOMELAND SECURITY PRESIDENTIAL DIRECTIVE /HSPD 25 (JANUARY 9, 2009).

January 9, 2009

NATIONAL SECURITY PRESIDENTIAL DIRECTIVE/NSPD -- 66
HOMELAND SECURITY PRESIDENTIAL DIRECTIVE/HSPD -- 25

MEMORANDUM FOR THE VICE PRESIDENT
THE SECRETARY OF STATE
THE SECRETARY OF THE TREASURY
THE SECRETARY OF DEFENSE
THE ATTORNEY GENERAL
THE SECRETARY OF THE INTERIOR
THE SECRETARY OF COMMERCE
THE SECRETARY OF HEALTH AND HUMAN SERVICES
THE SECRETARY OF TRANSPORTATION
THE SECRETARY OF ENERGY
THE SECRETARY OF HOMELAND SECURITY
ASSISTANT TO THE PRESIDENT AND CHIEF OF STAFF
ADMINISTRATOR OF THE ENVIRONMENTAL PROTECTION AGENCY
DIRECTOR OF THE OFFICE OF MANAGEMENT AND BUDGET
DIRECTOR OF NATIONAL INTELLIGENCE
ASSISTANT TO THE PRESIDENT FOR NATIONAL SECURITY AFFAIRS
COUNSEL TO THE PRESIDENT
ASSISTANT TO THE PRESIDENT AND DEPUTY NATIONAL SECURITY ADVISOR FOR INTERNATIONAL ECONOMIC AFFAIRS
ASSISTANT TO THE PRESIDENT FOR HOMELAND SECURITY AND COUNTERTERRORISM
CHAIRMAN, COUNCIL ON ENVIRONMENTAL QUALITY
DIRECTOR OF THE OFFICE OF SCIENCE AND TECHNOLOGY POLICY
CHAIRMAN OF THE JOINT CHIEFS OF STAFF
COMMANDANT, U.S. COAST GUARD
DIRECTOR, NATIONAL SCIENCE FOUNDATION

SUBJECT: Arctic Region Policy

I. PURPOSE

A. This directive establishes the policy of the United States with respect to the Arctic region and directs related implementation actions. This directive supersedes Presidential Decision Directive/NSC-26 (PDD-26; issued 1994) with respect to Arctic policy but not Antarctic policy; PDD-26 remains in effect for Antarctic policy only.

B. This directive shall be implemented in a manner consistent with the Constitution and laws of the United States, with the obligations of the United States under the treaties and other international agreements to which the United States is a party, and with customary international law as recognized by the United States, including with respect to the law of the sea.

II. BACKGROUND

A. The United States is an Arctic nation, with varied and compelling interests in that region. This directive takes into account several developments, including, among others:

1. Altered national policies on homeland security and defense;
2. The effects of climate change and increasing human activity in the Arctic region;
3. The establishment and ongoing work of the Arctic Council; and
4. A growing awareness that the Arctic region is both fragile and rich in resources.

III. POLICY

A. It is the policy of the United States to:

1. Meet national security and homeland security needs relevant to the Arctic region;
2. Protect the Arctic environment and conserve its biological resources;
3. Ensure that natural resource management and economic development in the region are environmentally sustainable;
4. Strengthen institutions for cooperation among the eight Arctic nations (the United States, Canada, Denmark, Finland, Iceland, Norway, the Russian Federation, and Sweden);
5. Involve the Arctic's indigenous communities in decisions that affect them; and
6. Enhance scientific monitoring and research into local, regional, and global environmental issues.

B. National Security and Homeland Security Interests in the Arctic

1. The United States has broad and fundamental national security interests in the Arctic region and is prepared to operate either independently or in conjunction with other states to safeguard these interests. These interests include such matters as missile defense and early warning; deployment of sea and air systems for strategic sealift, strategic deterrence, maritime presence, and maritime security operations; and ensuring freedom of navigation and overflight.
2. The United States also has fundamental homeland security interests in preventing terrorist attacks and mitigating those criminal or hostile acts that could increase the United States vulnerability to terrorism in the Arctic region.
3. The Arctic region is primarily a maritime domain; as such, existing policies and authorities relating to maritime areas continue to apply, including those relating to law enforcement.[1] Human activity in the Arctic region is increasing and is projected to increase further in coming years. This requires the United States to assert a more active and influential national presence to protect its Arctic interests and to project sea power throughout the region.
4. The United States exercises authority in accordance with lawful claims of United States sovereignty, sovereign rights, and jurisdiction in the Arctic region, including sovereignty within the territorial sea, sovereign rights and jurisdiction within the United States exclusive economic zone and on the continental shelf, and appropriate control in the United States contiguous zone.
5. Freedom of the seas is a top national priority. The Northwest Passage is a strait used for international navigation, and the Northern Sea Route includes straits used for international navigation; the regime of transit passage applies to passage through those

straits. Preserving the rights and duties relating to navigation and overflight in the Arctic region supports our ability to exercise these rights throughout the world, including through strategic straits.

6. <u>Implementation</u>: In carrying out this policy as it relates to national security and homeland security interests in the Arctic, the Secretaries of State, Defense, and Homeland Security, in coordination with heads of other relevant executive departments and agencies, shall:

   a. Develop greater capabilities and capacity, as necessary, to protect United States air, land, and sea borders in the Arctic region;

   b. Increase Arctic maritime domain awareness in order to protect maritime commerce, critical infrastructure, and key resources;

   c. Preserve the global mobility of United States military and civilian vessels and aircraft throughout the Arctic region;

   d. Project a sovereign United States maritime presence in the Arctic in support of essential United States interests; and

   e. Encourage the peaceful resolution of disputes in the Arctic region.

C. International Governance

1. The United States participates in a variety of fora, international organizations, and bilateral contacts that promote United States interests in the Arctic. These include the Arctic Council, the International Maritime Organization (IMO), wildlife conservation and management agreements, and many other mechanisms. As the Arctic changes and human activity in the region increases, the United States and other governments should consider, as appropriate, new international arrangements or enhancements to existing arrangements.

2. The Arctic Council has produced positive results for the United States by working within its limited mandate of environmental protection and sustainable development. Its subsidiary bodies, with help from many United States agencies, have developed and undertaken projects on a wide range of topics. The Council also provides a beneficial venue for interaction with indigenous groups. It is the position of the United States that the Arctic Council should remain a high-level forum devoted to issues within its current mandate and not be transformed into a formal international organization, particularly one with assessed contributions. The United States is nevertheless open to updating the structure of the Council, including consolidation of, or making operational changes to, its subsidiary bodies, to the extent such changes can clearly improve the Council's work and are consistent with the general mandate of the Council.

3. The geopolitical circumstances of the Arctic region differ sufficiently from those of the Antarctic region such that an "Arctic Treaty" of broad scope -- along the lines of the Antarctic Treaty -- is not appropriate or necessary.

4. The Senate should act favorably on U.S. accession to the U.N. Convention on the Law of the Sea promptly, to protect and advance U.S. interests, including with respect to the Arctic. Joining will serve the national security interests of the United States, including the maritime mobility of our Armed Forces worldwide. It will secure U.S. sovereign rights over extensive marine areas, including the valuable natural resources they contain. Accession will promote U.S. interests in the environmental health of the oceans. And it will give the United States a seat at the table when the rights that are vital to our interests are debated and interpreted.

5. <u>Implementation</u>: In carrying out this policy as it relates to international governance, the Secretary of State, in coordination with heads of other relevant executive departments and agencies, shall:

   a. Continue to cooperate with other countries on Arctic issues through the United Nations (U.N.) and its specialized agencies, as well as through treaties such as the U.N. Framework Convention on Climate Change, the Convention on International Trade in Endangered Species of Wild Fauna and Flora, the Convention on Long Range Transboundary Air Pollution and its protocols, and the Montreal Protocol on Substances that Deplete the Ozone Layer;

   b. Consider, as appropriate, new or enhanced international arrangements for the Arctic to address issues likely to arise from expected increases in human activity in that region, including shipping, local development and subsistence, exploitation of living marine resources, development of energy and other resources, and tourism;

   c. Review Arctic Council policy recommendations developed within the ambit of the Council's scientific reviews and ensure the policy recommendations are subject to review by Arctic governments; and

   d. Continue to seek advice and consent of the United States Senate to accede to the 1982 Law of the Sea Convention.

D. **Extended Continental Shelf and Boundary Issues**

1. Defining with certainty the area of the Arctic seabed and subsoil in which the United States may exercise its sovereign rights over natural resources such as oil, natural gas, methane hydrates, minerals, and living marine species is critical to our national interests in energy security, resource management, and environmental protection. The most effective way to achieve international recognition and legal certainty for our extended continental shelf is through the procedure available to States Parties to the U.N. Convention on the Law of the Sea.

2. The United States and Canada have an unresolved boundary in the Beaufort Sea. United States policy recognizes a boundary in this area based on equidistance. The United States recognizes that the boundary area may contain oil, natural gas, and other resources.

3. The United States and Russia are abiding by the terms of a maritime boundary treaty concluded in 1990, pending its entry into force. The United States is prepared to enter the agreement into force once ratified by the Russian Federation.

4. <u>Implementation</u>: In carrying out this policy as it relates to extended continental shelf and boundary issues, the Secretary of State, in coordination with heads of other relevant executive departments and agencies, shall:

   a. Take all actions necessary to establish the outer limit of the continental shelf appertaining to the United States, in the Arctic and in other regions, to the fullest extent permitted under international law;

   b. Consider the conservation and management of natural resources during the process of delimiting the extended continental shelf; and

   c. Continue to urge the Russian Federation to ratify the 1990 United States-Russia maritime boundary agreement.

E. **Promoting International Scientific Cooperation**

1. Scientific research is vital for the promotion of United States interests in the Arctic region. Successful conduct of U.S. research in the Arctic region requires access throughout the Arctic Ocean and to terrestrial sites, as well as viable international mechanisms for sharing access to research platforms and timely exchange of samples, data, and analyses. Better coordination with the Russian Federation, facilitating access to its domain, is particularly important.

2. The United States promotes the sharing of Arctic research platforms with other countries in support of collaborative research that advances fundamental understanding of the Arctic region in general and potential Arctic change in particular. This could include collaboration with bodies such as the Nordic Council and the European Polar Consortium, as well as with individual nations.

3. Accurate prediction of future environmental and climate change on a regional basis, and the delivery of near real-time information to end-users, requires obtaining, analyzing, and disseminating accurate data from the entire Arctic region, including both paleoclimatic data and observational data. The United States has made significant investments in the infrastructure needed to collect environmental data in the Arctic region, including the establishment of portions of an Arctic circumpolar observing network through a partnership among United States agencies, academic collaborators, and Arctic residents. The United States promotes active involvement of all Arctic nations in these efforts in order to advance scientific understanding that could provide the basis for assessing future impacts and proposed response strategies.

4. United States platforms capable of supporting forefront research in the Arctic Ocean, including portions expected to be ice-covered for the foreseeable future, as well as seasonally ice-free regions, should work with those of other nations through the establishment of an Arctic circumpolar observing network. All Arctic nations are members of the Group on Earth Observations partnership, which provides a framework for organizing an international approach to environmental observations in the region. In addition, the United States recognizes that academic and research institutions are vital partners in promoting and conducting Arctic research.

5. Implementation: In carrying out this policy as it relates to promoting scientific international cooperation, the Secretaries of State, the Interior, and Commerce and the Director of the National Science Foundation, in coordination with heads of other relevant executive departments and agencies, shall:
   a. Continue to play a leadership role in research throughout the Arctic region;
   b. Actively promote full and appropriate access by scientists to Arctic research sites through bilateral and multilateral measures and by other means;
   c. Lead the effort to establish an effective Arctic circumpolar observing network with broad partnership from other relevant nations;
   d. Promote regular meetings of Arctic science ministers or research council heads to share information concerning scientific research opportunities and to improve coordination of international Arctic research programs;
   e. Work with the Interagency Arctic Research Policy Committee (IARPC) to promote research that is strategically linked to U.S. policies articulated in this directive, with input from the Arctic Research Commission; and
   f. Strengthen partnerships with academic and research institutions and build upon the relationships these institutions have with their counterparts in other nations.

F. **Maritime Transportation in the Arctic Region**

1. The United States priorities for maritime transportation in the Arctic region are:
   a. To facilitate safe, secure, and reliable navigation;
   b. To protect maritime commerce; and
   c. To protect the environment.
2. Safe, secure, and environmentally sound maritime commerce in the Arctic region depends on infrastructure to support shipping activity, search and rescue capabilities, short- and long-range aids to navigation, high-risk area vessel-traffic management, iceberg warnings and other sea ice information, effective shipping standards, and measures to protect the marine environment. In addition, effective search and rescue in the Arctic will require local, State, Federal, tribal, commercial, volunteer, scientific, and multinational cooperation.
3. Working through the International Maritime Organization (IMO), the United States promotes strengthening existing measures and, as necessary, developing new measures to improve the safety and security of maritime transportation, as well as to protect the marine environment in the Arctic region. These measures may include ship routing and reporting systems, such as traffic separation and vessel traffic management schemes in Arctic chokepoints; updating and strengthening of the Guidelines for Ships Operating in Arctic Ice-Covered Waters; underwater noise standards for commercial shipping; a review of shipping insurance issues; oil and other hazardous material pollution response agreements; and environmental standards.
4. Implementation: In carrying out this policy as it relates to maritime transportation in the Arctic region, the Secretaries of State, Defense, Transportation, Commerce, and Homeland Security, in coordination with heads of other relevant executive departments and agencies, shall:
   a. Develop additional measures, in cooperation with other nations, to address issues that are likely to arise from expected increases in shipping into, out of, and through the Arctic region;
   b. Commensurate with the level of human activity in the region, establish a risk-based capability to address hazards in the Arctic environment. Such efforts shall advance work on pollution prevention and response standards; determine basing and logistics support requirements, including necessary airlift and icebreaking capabilities; and improve plans and cooperative agreements for search and rescue;
   c. Develop Arctic waterways management regimes in accordance with accepted international standards, including vessel traffic-monitoring and routing; safe navigation standards; accurate and standardized charts; and accurate and timely environmental and navigational information; and
   d. Evaluate the feasibility of using access through the Arctic for strategic sealift and humanitarian aid and disaster relief.

G. Economic Issues, Including Energy

1. Sustainable development in the Arctic region poses particular challenges. Stakeholder input will inform key decisions as the United States seeks to promote economic and energy security. Climate change and other factors are significantly affecting the lives of Arctic inhabitants, particularly indigenous communities. The United States affirms the importance to Arctic communities of adapting to climate change, given their particular vulnerabilities.

2. Energy development in the Arctic region will play an important role in meeting growing global energy demand as the area is thought to contain a substantial portion of the world's undiscovered energy resources. The United States seeks to ensure that energy development throughout the Arctic occurs in an environmentally sound manner, taking into account the interests of indigenous and local communities, as well as open and transparent market principles. The United States seeks to balance access to, and development of, energy and other natural resources with the protection of the Arctic environment by ensuring that continental shelf resources are managed in a responsible manner and by continuing to work closely with other Arctic nations.

3. The United States recognizes the value and effectiveness of existing fora, such as the Arctic Council, the International Regulators Forum, and the International Standards Organization.

4. Implementation: In carrying out this policy as it relates to economic issues, including energy, the Secretaries of State, the Interior, Commerce, and Energy, in coordination with heads of other relevant executive departments and agencies, shall:

   a. Seek to increase efforts, including those in the Arctic Council, to study changing climate conditions, with a view to preserving and enhancing economic opportunity in the Arctic region. Such efforts shall include inventories and assessments of villages, indigenous communities, subsistence opportunities, public facilities, infrastructure, oil and gas development projects, alternative energy development opportunities, forestry, cultural and other sites, living marine resources, and other elements of the Arctic's socioeconomic composition;

   b. Work with other Arctic nations to ensure that hydrocarbon and other development in the Arctic region is carried out in accordance with accepted best practices and internationally recognized standards and the 2006 Group of Eight (G-8) Global Energy Security Principles;

   c. Consult with other Arctic nations to discuss issues related to exploration, production, environmental and socioeconomic impacts, including drilling conduct, facility sharing, the sharing of environmental data, impact assessments, compatible monitoring programs, and reservoir management in areas with potentially shared resources;

   d. Protect United States interests with respect to hydrocarbon reservoirs that may overlap boundaries to mitigate adverse environmental and economic consequences related to their development;

   e. Identify opportunities for international cooperation on methane hydrate issues, North Slope hydrology, and other matters;

   f. Explore whether there is a need for additional fora for informing decisions on hydrocarbon leasing, exploration, development, production, and transportation, as well as shared support activities, including infrastructure projects; and

   g. Continue to emphasize cooperative mechanisms with nations operating in the region to address shared concerns, recognizing that most known Arctic oil and gas resources are located outside of United States jurisdiction.

H. Environmental Protection and Conservation of Natural Resources

1. The Arctic environment is unique and changing. Increased human activity is expected to bring additional stressors to the Arctic environment, with potentially serious consequences for Arctic communities and ecosystems.

2. Despite a growing body of research, the Arctic environment remains poorly understood. Sea ice and glaciers are in retreat. Permafrost is thawing and coasts are eroding. Pollutants from within and outside the Arctic are contaminating the region. Basic data are lacking in many fields. High levels of uncertainty remain concerning the effects of climate change and increased human activity in the Arctic. Given the need for decisions to be based on sound scientific and socioeconomic information, Arctic environmental research, monitoring, and vulnerability assessments are top priorities. For example, an understanding of the probable consequences of global climate variability and change on Arctic ecosystems is essential to guide the effective long-term management of Arctic natural resources and to address socioeconomic impacts of changing patterns in the use of natural resources.

3. Taking into account the limitations in existing data, United States efforts to protect the Arctic environment and to conserve its natural resources must be risk-based and proceed on the basis of the best available information.

4. The United States supports the application in the Arctic region of the general principles of international fisheries management outlined in the 1995 Agreement for the Implementation of the Provisions of the United Nations Convention on the Law of the Sea of December 10, 1982, relating to the Conservation and Management of Straddling Fish Stocks and Highly Migratory Fish Stocks and similar instruments. The United States endorses the protection of vulnerable marine ecosystems in the Arctic from destructive fishing practices and seeks to ensure an adequate enforcement presence to safeguard Arctic living marine resources.

5. With temperature increases in the Arctic region, contaminants currently locked in the ice and soils will be released into the air, water, and land. This trend, along with increased human activity within and below the Arctic, will result in increased introduction of contaminants into the Arctic, including both persistent pollutants (e.g., persistent organic pollutants and mercury) and airborne pollutants (e.g., soot).

6. Implementation: In carrying out this policy as it relates to environmental protection and conservation of natural resources, the Secretaries of State, the Interior, Commerce, and Homeland Security and the Administrator of the Environmental Protection Agency, in coordination with heads of other relevant executive departments and agencies, shall:
    a. In cooperation with other nations, respond effectively to increased pollutants and other environmental challenges;
    b. Continue to identify ways to conserve, protect, and sustainably manage Arctic species and ensure adequate enforcement presence to safeguard living marine resources, taking account of the changing ranges or distribution of some species in the Arctic. For species whose range includes areas both within and beyond United States jurisdiction, the United States shall continue to collaborate with other governments to ensure effective conservation and management;
    c. Seek to develop ways to address changing and expanding commercial fisheries in the Arctic, including through consideration of international agreements or organizations to govern future Arctic fisheries;
    d. Pursue marine ecosystem-based management in the Arctic; and
    e. Intensify efforts to develop scientific information on the adverse effects of pollutants on human health and the environment and work with other nations to reduce the introduction of key pollutants into the Arctic.

IV. Resources and Assets

A. Implementing a number of the policy elements directed above will require appropriate resources and assets. These elements shall be implemented consistent with applicable law and authorities of agencies, or heads of agencies, vested by law, and subject to the availability of appropriations. The heads of executive departments and agencies with responsibilities relating to the Arctic region shall work to identify future budget, administrative, personnel, or legislative proposal requirements to implement the elements of this directive.

GEORGE W. BUSH

# # #

---

[1] These policies and authorities include Freedom of Navigation (PDD/NSC-32), the U.S. Policy on Protecting the Ocean Environment (PDD/NSC-36), Maritime Security Policy (NSPD-41/HSPD-13), and the National Strategy for Maritime Security (NSMS).

Source: The White House

BIBLIOGRAPHY

Alaska state Hospital and Nursing Home Association. http://www.ashnha.com/about/contact-us/ (accessed March 16, 2013).

Aldo Chircop. "The Growth of International Shipping in the Arctic: Is a Regulatory Review Timely?" *The International Journal of Marine and Coastal Law*, 24, no. 2 (2009), 355-380.

AMAP Assessment 2009: *Human Health in the Arctic*. Arctic Monitoring and Assessment Programme (AMAP), Oslo, 2009.

Anderson, Alun. *After the Ice: Life, Death, and Geopolitics in the New Arctic*. Smithsonian: HarperCollins, 2009.

Arctic Council. *Arctic Marine Shipping Assessment 2009 Report*, April 2009.

Arctic Council. *Declaration on the Establishment of the Arctic Council* (Ottawa, 1996).

Arctic Council. *Sustainable Development Program (SD Program)*, http://www.sdwg.org/content.php?doc=12 (accessed 16 Mar 2013).

Arctic Council. *The Ilulissat Declaration* (Ilulissat, Greenland, 2008).

Berkman, Paul Arthur and Royal United Services Institute for Defence and Security Studies. *Environmental Security in the Arctic Ocean: Promoting Co-cooperation and Preventing Conflict*. Whitehall paper, 2010.

Borgerson, Scott G. "Arctic Meltdown: The Economic and Security Implications of Global Warming," *Foreign Affairs*, vol 87, no. 2. March/April 2008, 69.

Brigham, Lawson W. CAPT, USCG (ret). "The Fast-Changing Maritime Arctic," *Proceedings*, U.S. Naval Institute (May 2010), 58.

Bureau of Ocean Energy Management. "2012- 2017 Lease Sale Schedule", http://www.boem.gov/Oil-and-Gas-Energy-Program/Leasing/Five-Year-Program/Lease-Sale-Schedule/2012---2017-Lease-Sale-Schedule.aspx (accessed March 17, 2013).

Bureau of Ocean Energy Management. "Five Year Outer Continental Shelf (OCS) Oil and Gas Leasing Program," http://www.boem.gov/5-year/2012-2017/ (accessed March 17, 2013).

Bureau of Ocean Energy Management. "Shell 2012 Exploration Plan – Chukchi Sea," http://www.boem.gov/ShellChukchi2012/ (accessed March 17, 2013).

Byers, Michael. *Who Owns the Arctic? Understanding Sovereignty Disputes in the North*. Vancouver: Douglas & McIntyre Publishers Inc., 2009.

Cacas, Max. "Coast Guard Prepares as Arctic Region Heats Up" *Signal,* June 2012, 40.

Campbell, Megan L. "*United States Arctic Ocean Management and the Law of the Sea Convention*," (externship paper, U.S. Dept. of Commerce, National Oceanic and Atmospheric Administration, September 2008.

Canada, Minister of Fisheries and Oceans Canada. Canadian Coast Guard. *Ice Navigation in Canadian Waters, 2012*. Ottawa, Ontario, 2012.

"Canada and United States of America Agreement on Arctic Cooperation," January 11, 1988, *United Nations Treaty Series*, no. 1852-i-31529, (Ottawa, 1988).

Currie, Duncan E.J. "Sovereignty and Conflict in the Arctic Due to Climate Change: Climate Change and the Legal Status of the Arctic Ocean," *Global Law.com* (August 5, 2007) http://www.globelaw.com/LawSea/Arctic%20claims%20and%20climate%20change.pdf (accessed 12 Sept 2012).

Draft Arctic Council Agreement as released by Greenpeace. "Agreement on Cooperation on Marine Oil Pollution Preparedness and Response in the Arctic," 11 October 2012 http://www.greenpeace.org/international/Global/international/briefings/climate/ArcticCouncilDra ftAgreementCooperationOPPR.pdf (accessed 6 Mar 2013).

Emmerson, Charles. *The Future History of the Arctic.* New York: Public Affairs, 2010.

Environmental Law Institute. "David A. Bolton Biography," http://www.eli.org/ pdf/ocean/seminars/balton_bio.pdf (accessed Feb 4, 2013).

Extended Continental Shelf Project. "2010 Extended Continental Shelf Survey," http://continentalshelf.gov/missions/10arctic/welcome.html (accessed March 17, 2013).

Extended Continental Shelf Project. "Extended Continental Shelf Summary of Missions," http://continentalshelf.gov/missions.html (accessed February 3, 2013).

Gourley, Julia L. "Pressing Issues in the Arctic" (Department of State PowerPoint Presentation given September, 7, 2011, Anchorage, AK) http://onlinepubs.trb.org/onlinepubs/mb/ 2011Fall/ppt/1gourley.pdf (accessed Feb 4, 2013).

Grant, Shelagh D. *Polar Imperative: A History of Arctic Sovereignty in North America.* Vancouver: Douglas and McIntyre, 2010.

Heygster, Georg. *"Arctic Sea Ice Extent Small as Never Before",* University of Bremen, http://www.iup.uni-bremen.de:8084/amsr/minimum2011-en.pdf (accessed 17 January 2013.

Interagency Arctic Research Policy Committee. *Arctic Research Plan: FY2013-2017,* September, 2012.

International Arctic Buoy Programme, Polar Science Center, University of Washington, "Overview," http://iabp.apl.washington.edu/index.html (accessed February 11, 2013).

International Arctic Systems for Observing the Atmosphere, Tiksi, Russia, "Tisksi International Hydrometeorological Observatory," http://iasoa.org/iasoa/index.php?option= com_content& task=view&id=81&Itemid=119 (accessed February 3, 2013).

International Maritime Organization. "Introduction to IMO," http://www.imo.org/About/ Pages/Default.aspx (accessed January 3, 2013).

International Maritime Organization. *Convention on the International Regulations for Preventing Collisions at Sea, 1972,* codified at 33 USCS § 1602.

Jolin, Dan. "Admiral Confident in Coast Guard Arctic Readiness," *Juneau Empire* (August 9, 2012), http://juneauempire.com/state/2012-08-09/admiral-confident-coast-guard-Arctic-readiness (accessed September 11, 2012).

Kotter, John and Rathgeber, Holger. *Our Iceberg is Melting* (New York: St Martin's Press, 2005).

Marine Mammal Protection Act (MMPA), U.S. Code 16§1361.

Marport. Happenings from the World of Deep Sea Technology, "United Stated Arctic Fishing Policy Latest in Can-Am Dispute," September 3, 2009, http://blog.marport.com/2009/09/03/1329/ (accessed March 9, 2013).

Murphy, Kim. "Coast Guard Beefs Up Deployment in the U.S. Arctic," *Los Angeles Times* (March 01, 2012) http://articles.latimes.com/2012/mar/01/nation/la-na-nn-coast-guard-Arctic-20120301 (accessed September 11, 2012).

National Oceanic and Atmospheric Administration. "Arctic Navigation; How is NOAA improving the nautical charts ships need?" http://oceanservice.noaa.gov/ economy/arctic/#4 (accessed Mar 8, 2013).

National Oceanic and Atmospheric Administration. "Arctic Navigation; What about hydrographic surveying for navigation?" http://oceanservice.noaa.gov/economy/arctic/#5 (accessed Mar 8, 2013).

National Oceanic and Atmospheric Administration. "Research institutions and organizations focused on the Arctic," http://www.arctic.noaa.gov/orgs.html (accessed February 3, 1013).

National Oceanic and Atmospheric Administration. Ocean Explorer, "Russian-U.S. Arctic Census 2012," http://oceanexplorer.noaa.gov/explorations/12arctic/welcome.html (accessed March 19, 2013).

National Research Council. *Polar Icebreakers in a Changing World: An Assessment of U.S. Needs*, National Academy of Sciences (Washington, DC, 2006), 1 and 4.

National Science Foundation. "Arctic Research and Education," http://www.nsf.gov/funding/ pgm_summ.jsp?pims_id=13448 (accessed February 3, 2013).

National Science Foundation. "Interagency Arctic Research Policy Committee (IARPC) – Arctic Research Plan," http://www.nsf.gov/od/opp/arctic/iarpc/arc_res_plan_index.jsp (accessed March 19, 2013).

National Science Foundation. "Interagency Arctic Research Policy Committee Principals 2012," IARPC Principals list, http://www.nsf.gov/od/opp/arctic/iarpc/iarpc_principals2012.jsp (accessed February 3, 2013).

National Science Foundation. Arctic Observing Network (AON), "Program Guidelines,' http://www.nsf.gov/funding/pgm_summ.jsp?pims_id=503222&org=NSF (accessed February 12, 2013).

National Science Foundation. Fact Sheet, "The Arctic Observing Network (AON)" July 10, 2007. http://www.nsf.gov/news/news_summ.jsp?cntn_id=109687 (accessed March 19, 2013).

North Pacific Fishery Management Council. "Fishery Management Plan for Fish Resources of the Arctic Management Area", Anchorage, Alaska, 2009. ES-4. https://alaskafisheries.noaa. gov/npfmc/ PDFdocuments/fmp/Arctic/ArcticFMP.pdf (accessed March 18, 2013).

Northwest Territorial Mint. "Palladium Uses," http://bullion.nwtmint.com/palladium_uses.php (accessed April 1, 2013).

Petersen, Luke R. "International Striat or Internal Waters? : The navigational potential of the Northwest Passage," *USCG Proceedings* (Summer 2009), 45.

President of the United States. *National Security Presidential Directive 66/National Homeland Security Presidential Directive 25.* January 9, 2009.

President of the United States. *National Security Strategy.* Washington, DC. May 2010.

President, Proclamation. "Fisheries of the United States Exclusive Economic Zone Off Alaska; Fisheries of the Arctic Management Area; Bering Sea Subarea," *Federal Register* 74, no. 211, November 3, 2009, 56734.

Study of Environmental Arctic Change (SEARCH). "Development of SEARCH," "1990-1999: Early Activities," http://www.arcus.org/search/sciencecoordination/development (accessed February 3, 2013).

Study of Environmental Arctic Change. "2012 U.S. Arctic Observing Coordination Workshop," http://www.arcus.org/search/aon (accesses Feb 15, 2013).

Submarine Force U.S. Pacific Fleet. "Arctic Submarine Laboratory," http://www.csp.navy.mil/ asl/index.htm (accessed February 5, 2013).

Sustaining Arctic Observing Networks (SAON). "Board Meetings," http://www.arcticobserving.org/news (accessed February 3, 2013).

Sustaining Arctic Observing Networks. National Reports, "US National Report," http://www.arcticobserving.org/national-reports (accessed February 3, 2013).

Title 33. codified at *U.S. Code* 33 (2003), § 164046.

Towell, Pat and Else, Daniel H. CRS Report. "Defense: FY2013 Authorization and Appropriations." September 5, 2012. Washington, DC, 1-50.

Troedsson, Peter, Capt USCG. "Leading Preparedness for an International Oil Spill Response in the Arctic," *The Fletcher Forum of World Affairs* (February 2013) http://www.fletcherforum.org/2013/02/20/troedsson/ (accessed 6 Mar 2013).

U.S. Arctic Research Commission. "About USARC," http://www.arctic.gov/about.html (accessed March 19, 2013).

U.S. Arctic Research Commission. *Report on the Goals and Objectives for Arctic Research 2013– 2014,* (Arlington, VA. February 2013), 1-3.

U.S. Coast Guard. Publication 1, *U.S. Coast Guard: America's Maritime Guardian.* May 2009

U.S. Coast Guard. Publication 3-0, *Operations.* February 2012

U.S. Coast Guard. Coast Guard District 17 External Affairs Office news release, "Imagery Available: Coast Guard completes Arctic Shield 2012," 1 Nov 2012, http://www.uscgnews.com /go/doc/4007/1594651/.

U.S. Coast Guard. Commandant Instruction 16003.1, *U.S. Coast Guard Arctic Strategic Approach,* (Washington, DC, April 26, 2011).

U.S. Coast Guard. National Response Center "National Response System," http://www.nrc.uscg.mil/nrsinfo.html (accessed Feb 4, 2013).

U.S. Coast Guard. *Ocean Guardian,* September 20, 2004. Washington, DC.1

U.S. Coast Guard. "USCGC POLAR SEA (WAGB 11)," http://www.uscg.mil/pacarea/ cgcpolarsea/; and U.S. Coast Guard "CGC Healy Ship's Characteristics," http://www.uscg.mil/pacarea/ cgchealy/ship.asp (accessed Feb 9, 2013).

U.S. Coast Guard. "USCG District 17 Arctic Brief" Power Point Slides by Rear Admiral Christopher C. Colvin, Commander 17th Coast Guard District, January 27, 2011.

U.S. Coast Guard. *Coast Guard Commandant Admiral Bob Papp's State of the Coast Guard Address.* February 23, 2012.

U.S. Coast Guard. *U.S. Coast Guard 2012 Posture Statement with 2013 Budget in Brief.* February, 2012.

U.S. Congress 112th 2nd session. HR 2838. Appropriations for the Coast Guard for fiscal year 2013 through 2014, 42.

U.S. Congress. House Committee on Appropriations. *Statement of Admiral Jams A. Winnefeld, Jr, U.S. Navy, Commander U.S. Northern Command and North American Aerospace Defense Command:*

Hearing before the House Armed Services Committee. 112 Cong. 2nd sess., March 6, 2012.

U.S. Congress. House Committee on Appropriations. *Testimony of Admiral Robert Papp Commandant, U.S. Coast Guard on "USCG FY 2013 Budget": Hearing before the Subcommittee on Homeland Security*. 112 Cong. 2nd sess., March 6, 2012.

U.S. Congress. Senate Appropriations Subcommittee. *Testimony of Admiral Robert Papp Commandant, U.S. Coast Guard on Coast Guard Operations in the Arctic: Hearing before the Subcommittee on Appropriation*. Given in Kodiak, AK. August 6, 2012.

U.S. Congress. Senate Committee on Appropriations. *Testimony of Admiral Robert Papp Commandant, U.S. Coast Guard on "USCG FY 2013 Budget": Hearing before the Subcommittee on Homeland Security*. 112 Cong. 2nd sess., May 9, 2012.

U.S. Congress. Senate Committee on Commerce, Science, and Transportation. *Testimony of Admiral Robert Papp Commandant, U.S. Coast Guard on "Arctic Operations": Hearing before the Subcommittee on Oceans, Atmosphere, Fisheries, and Coast Guard*. 112 Cong. 1st sess., July 27, 2011.

U.S. Congress. Senate Committee on Commerce, Science, and Transportation. *Testimony of Admiral Robert Papp Commandant, U.S. Coast Guard on "Arctic Operations": Hearing before the Subcommittee on Oceans, Atmosphere, Fisheries, and Coast Guard*. 112 Cong. 1st sess., June 23, 2011.

U.S. Congress. Senate. *A joint resolution directing the United States to initiate international discussions and take necessary steps with other nations to negotiate an agreement for managing migratory and transboundary fish stocks in the Arctic Ocean*. S. 17. 110th Cong., 2d sess., May 23, 2008: 2.

U.S. Congressional Research Service. *Changes in the Arctic: Background and Issues for Congress* by Ronald O'Rourke. Washington, DC, 2013.

U.S. Congressional Research Service. *Coast Guard, Polar Icebreaker Modernization Background, Issues, and Options for Congress* by Ronald O'Rourke. Washington, DC, 2008.

U.S. Congressional Research Service. *Changes in the Arctic: Background and Issues for Congress*. Ronald O'Rourke, Coordinator. Washington, DC, August 1, 2012.

U.S. Department of Commerce. *Budget in Brief Fiscal Year 2013*, John E. Bryson, Secretary Washington, DC, 6.

U.S. Department of Commerce. "Memorandum of Understanding between the National Oceanic and Atmospheric Administration of the Department of Commerce of the United States of America and the Russian Academy of Sciences of the Russian Federation on the Cooperation in the Area of the World Oceans and Polar Regions." 2003.

U.S. Department of Commerce. National Oceanic and Atmospheric Administration, "Notice of Availability of a Draft Environmental Impact Statement for the Effects of Oil and Gas Activities in the Arctic Ocean," *Federal Register* 76, no. 251, December 30, 2011, 82275.

U.S. Department of Commerce. Public Law 94-265 Magnuson-Stevens Fishery Conservation and Management Act. U.S. Department of Commerce, National Oceanic and Atmospheric Administration, National Marine Fisheries Service January 12, 2007, 63.

U.S. Department of Defense. *Report to Congress on Arctic Operation and the Northwest Passage*. (Washington, DC, May 2011), 3-4.

U.S. Department of Defense. *"Overview"* Office Of The Under Secretary Of Defense (Comptroller) /

Chief Financial Officer, February 2012.

U.S. Department of Defense. *Sustaining US Global Leadership: Priorities for 21ˢᵗ Century Defense.* Washington, DC. January 5, 2012, 4.

U.S. Department of Energy. "On the Frontiers of a New Energy Source,"(Washington, DC, May 2, 2012) http://energy.gov/articles/frontiers-new-energy-source (accessed March 17, 2013).

U.S. Department of Homeland Security. *"Secretary Napolitano Announces Fiscal Year 2013 Budget Request"* Washington, DC, February 13, 2012. http://www.dhs.gov/news/2012/02/13 /secretary-napolitano-announces-fiscal-year-2013-budget-request (accessed March 19, 2013).

U.S. Department of Homeland Security. *FY 13 Budget-in-Brief,* Washington, DC, 87. http://www.dhs.gov/xlibrary/assets/mgmt/dhs-budget-in-brief-fy2013.pdf (accessed March 19, 2013).

U.S. Department of Homeland Security. *Quadrennial Homeland Security Review Report,* Washington, DC, February 2010.

U.S. Department of Homeland Security. *Strategic Plan Fiscal Years 2012-2016,* Washington, DC, February 2012.

U.S. Department of Homeland Security. U.S. Coast Guard. *Written testimony of U.S. Coast Guard Commandant Admiral Robert Papp, Jr. "Coast Guard Fiscal Year 2013 Budget Request": Hearingbefore the House Subcommittee on Coast Guard and Maritime Transportation.* March 7, 2012.

U.S. Department of Homeland Security. U.S. Coast Guard. *Written testimony of U.S. Coast Guard Commandant Admiral Robert Papp, Jr. "The Law of the Sea Convention (Treaty Doc. 103-39): Perspectives from the U.S. Military": Hearing before the Senate Committee on Foreign Relations.* June 14, 2012.

U.S. Department of Homeland Security. *Northern Border Strategy*, Washington, DC, June 2012. 6.

U.S. Department of Interior. *Fiscal Year 2013 The Interior Budget in Brief,* Washington, DC February 2012, DO-15

U.S. Department of State. http://www.state.gov/documents/organization/183755.pdfExecutive Budget Summary *"Function 150 & Other International Programs Fiscal Year 2013"*, 58 and 81.

U.S. Department of State. "Arctic Search and Rescue," http://www.state.gov/e/oes/ ocns/opa/arc/c29382.htm (accessed March 14, 2013).

U.S. Department of State. "Defining the Limits of the U.S. Continental Shelf," http://www.state.gov/e/oes/continentalshelf/ (accessed March 19, 2013).

U.S. Department of State. "Ocean and Polar Affairs," http://www.state.gov/e/oes/ocns/opa/.

U.S. Department of State. "U.S. Relations with Canada," http://www.state.gov/r/pa/ei/bgn/2089.htm (accessed March 14, 2013).

U.S. Department of State. Treaties in Force: A List of Treaties and Other International Agreements of the United States in Force on January 1, 2012 available www.state.gov/ documents/organization/202293.pdf (accessed March 17, 2013).

U.S. Department of State. U.S.- Russia Bilateral Commission. http://www.state.gov/p/eur/ ci/rs/usrussiabilat/index.htm (accessed March 19, 2013).

U.S. Department of the Navy. *U.S. Navy Arctic Roadmap,* Washington, DC, October 2009.

U.S. Government Accountability Office. GAO-10-870. *Report to Congressional Requesters. Coast Guard, Efforts to Identify Arctic Requirements Are Ongoing, but More Communication about Agency Planning Efforts Would be Beneficial.* September 2010.

U.S. Government Accountability Office. GAO-12-180 *Report to Congressional Committees. Arctic Capabilities: DOD Addressed Many Specified Reporting Elements in Its 2011 "Arctic Report" but Should Take Step to Meet Near-and-Long term Needs.* January 2012.

U.S. Government Accountability Office. GAO-12-254T. *Testimony Before the Subcommittee on Coast Guard and Maritime Transportation, Committee on Transportation and Infrastructure, House of Representative. Coast Guard: Observations on Arctic Requirements, Icebreakers, and Coordination with Stakeholder.* December 1, 2011.

U.S. National Oceanic and Atmospheric Administration. *NOAA's Arctic Vision & Strategy* (Washington, DC, February 2011), 6.

U.S. National Oceanic and Atmospheric Administration. "NOAA Ship Fairweather sets sail to map areas of the Arctic," (Washington, DC, July 7, 2011) http://www.noaanews. noaa.gov/stories2011/20110707_fairweather.html (accessed March 17, 2013).

U.S. Navy. *Arctic Environmental Assessment and Outlook Report; In support of the Navy Arctic Roadmap; Action Item 5.7*, Washington, DC, 2011.

U.S. Navy. Chief of Naval Operations memo dated 21 May 2010, *Navy Strategic Objectives for the Arctic*, Washington, DC, 2010.

U.S. Navy, U.S. Marine Corps, U.S. Coast Guard. *Naval Operations Concept 2010: Implementing the Maritime Strategy*, Washington, DC, 2010.

U.S. Senate Committee on Foreign Relations. *The Law of the Sea Convention (Treaty Doc. 103-39): The U.S. National Security and Strategic Imperatives for Ratification.* 112th Cong., 2d sess., 2012.

United Nations Convention on the Law of the Sea (UNCLOS). December 10, 1982, *United Nation Treaty Series*, volume 1833, registration Number I-31363.

University of Maryland. Center for Environmental Science Chesapeake Biological Laboratory. "Bering Ecoystem Study (BEST) and Bering Sea Integrated Ecosystem Research Program (BSIERP)". http://arctic.cbl.umces.edu/ (accessed February 3, 2013).

Warner, Jon Trent LCDR, USCG. "Supporting the Arctic Marine Transportation System," *USCG Proceedings*, Summer 2011, 68.